NEWTON'S PHILOSOPHY OF NATURE

SELECTIONS FROM HIS WRITINGS

Edited and Arranged with Notes by
H. S. THAYER

Introduction by
JOHN HERMAN RANDALL, JR.

DOVER PUBLICATIONS, INC.
Mineola, New York

Bibliographical Note

This Dover edition, first published in 2005, is an unabridged republication of the work originally published by Hafner Publishing Company, New York, in 1953.

International Standard Book Number: 0-486-44593-3

Manufactured in the United States of America
Dover Publications, Inc., 31 East 2nd Street, Mineola, N.Y. 11501

Contents

v

Editor's Preface

It is a great pity that Newton is so little read, especially in an age that prides itself on being scientifically-minded. For nothing is less scientific than to overlook the fact that present ideas have past antecedents and that present triumphs in science, as in everything else, come only by way of the arduous preparations and labors by which men in the past laid the foundations for subsequent successes. Newton stands as one of the few great men in history who have directed and shaped the course of things present. Not the least of his influence is to have given modern physics its revolutionary character. For we are apt to forget that without Newton's spectacular achievements there would have been far less of a fertile source from which such a revolution could have sprung.

Besides the *Principia* and occasional appearances of the *Optics*, the writings of Newton on the whole have remained inaccessible to students of philosophy, science, and literature, as well as to the general reader. It is the aim of this book to fill, to some extent, this unfortunate lacuna; that is, to provide a wider representation of the interests, problems, and characteristically diverse philosophic levels and directions along which Newton's thoughts traveled. Once given the chance, there are few persons who could fail to be interested in or profit by meeting the greatest scientific mind of the seventeenth century. This book was begun with the encouragement and expert advice of Professor John Herman Randall, Jr.; it was also my good fortune to secure his consent to write the Introduction.

The various selections have been grouped under five general headings. While some classification of this sort is necessary, in some instances it may also result in a certain unavoidable artificiality. For example, when Newton begins to talk about gravity, he often concludes with God, and separation of the two topics does some violence to his thought. In important places, in both the text and notes to the selections, where such a connection might be

vii

missed, cross-references have been inserted. The selections from the *Optics* (i.e., the Questions from Book III) allow the reader to follow Newton's own arrangement and development of those thoughts which concerned him most deeply. In these Questions, it will be observed, almost all of the fundamental ideas which occupied Newton—and which make up the body of the selections in this book—appear again, but within the ordered sequence of the speculative reflections to which he gave such brilliant articulation.

I must assume full responsibility for the particular selections included, their arrangement, and the historical and explanatory Notes at the end of the book. Where it has seemed necessary to add a few words either to the text or as footnotes, the additions appear in brackets. Newton's punctuation and spelling have been altered here and there to conform to current usage. The Bibliography lists the original sources of the selections, as well as a number of valuable secondary sources.

I am grateful to the University of California Press for permission to use Florian Cajori's excellent revision of Motte's translation of the *Principia*. Cajori's historical appendix is of importance; his revision of the text, in rendering the now obsolete mathematical terminology into modern equivalents, makes the California edition indispensable to the modern reader of Newton's masterpiece. My thanks go also to Professor E. A. Burtt, who has kindly granted me permission to quote, in the note to the General Scholium, from his justly well-known and valuable book *The Metaphysical Foundations of Modern Physical Science*. I have learned much from L. T. More's important and informative biography of Newton.

I wish to acknowledge my indebtedness to Dr. Jason L. Saunders for his careful preliminary study and subsequent translation from the Latin of Newton's letter to Oldenburg on hypotheses. To Professors Ernest Nagel, Marjorie H. Nicolson, and Herbert W. Schneider, I am also grateful for valuable suggestions and advice.

H. S. THAYER

What Isaac Newton Started

Isaac Newton is not only by general acclaim the greatest scientific genius the English-speaking peoples have produced, and one of the half-dozen towering giants of the intellectual movement that has distinguished the modern world from all other societies. He also gave his name to an entire age, which is more than even Darwin could do for the Age of Evolution or Freud for our own Age of Anxiety. He lived at just the right moment to reap the harvest sown by several generations of scientific pioneers. By temperament and intellectual sympathies he was able to weave together the two main strands of seventeenth-century scientific thought: the mathematical rationalism of the Continental tradition (in England then as now entrenched at Cambridge); and the "physico-mathematical experimental learning" which the Royal Society cultivated. His achievement in synthesis caught the imagination of the early eighteenth century, and he came to stand as the symbol of a broadly conceived new "natural philosophy," or physical science.

After two centuries of battles fought in the name of warring theologies and church polities, most men were only too glad to welcome this new natural philosophy as a secular alternative to religious quarrels of which they had grown tired. Many wanted to forget theology and get down to business, especially that middle class which in Western Europe had been growing so rapidly in economic strength and was now making ready to take over political power as well, in the great revolutions of the end of the century. What the middle class needed was a new set of ideas to provide the intellectual leverage for dislodging the lingering feudal landlords and breaking the hold of the older social controls of industry, now grown restrictive. For them, "Newtonian science" furnished a "Nature" fully as effective as the earlier "will of God." It had, in fact, at last demonstrated what the will of God really was; and what it demonstrated was that the Divine Will had

ix

decreed a mechanism that worked automatically without further interference. No wonder that the social philosophies that endeavored to extend scientific methods to human affairs pointed to a similar autonomous order as the highest wisdom for conducting the life of man. Thus the Newtonian philosophy of nature was made into what a later jargon calls "the ideology of the bourgeois revolution."

This was the second time that Western society had turned to a secular and scientific body of ideas to consecrate its highest values and ground them in the nature of things—the first such episode in modern times. In the thirteenth century it had found in Aristotelian thought an admirable instrument for organizing its entire culture in the service of its traditional religious ideals. Now Newtonian concepts proved equally available for giving force and direction to aims that, if no less religious, were new and secular. Here was a new harmony of knowledge and aspiration, a new heavenly city buttressed by a new scientific truth.

It is a valid question whether the Aristotelian science that made central the understanding of life in general and of man's life in particular did not provide a more adequate human wisdom for the human animal than Newtonian mechanics, or indeed whether the Platonic science of the Hellenistic world did not do better than either in furnishing a spiritual wisdom for the spirit of man. But Newton himself, as well as those who went on eagerly to construct a new social wisdom on the basis of his philosophy of nature, and even those who attacked him and them because they preferred the older wisdom, would all alike have been amazed at the more recent contention that natural science has nothing to do with "values," that it can and should itself remain "value-free," and that those seeking a direction for human life have nothing to learn from our best knowledge of the nature of things. Newton would certainly have rejoiced that "Newtonian thought" signifies not only a necessary stage in the development of mathematical physics but also a gleaming pinnacle in the moral and religious life of Western culture. Newton's mechanics is no doubt a fragile reed on which to build a viable science of man and society. But who, amidst the voluntarisms and irrationalisms of the nineteenth century and after, can claim that we wiser and more sophisticated mortals have achieved more admirable ideals than the benevolence, tolerance, intellectual freedom, cosmopolitanism, and peace to which Newtonian scientific rationalism inspired so many in the eighteenth century? Even a little science—and that by our standards woefully inadequate for the human enterprise to which men so bravely applied it—is a thing of infinite promise for human values.

In our degenerate and more specialized days, those likely to turn to the reading of Newton will no doubt have divergent interests. The historian of science will find Newton a master builder of the foundations of that great edifice whose never-finished topmost stories are still being built into the clouds. He will be concerned with Newton's permanent achievements, with his methods and concepts, with all that Newton himself thought could be "deduced from phenomena" and hence belonged properly to "experimental philosophy" or natural science. The "hypotheses" which fascinated Newton himself, but went beyond those rigorous limits, may attract his attention for a moment, for some of them managed to gain a later accrediting. But he will smile at Newton's religious ideas and be apt to regard him as incomprehensibly schizophrenic. At best, he will grant that since Newton always knew himself when he was engaging in natural philosophy and when in theology, we can concentrate on the former and profitably neglect the latter.

This viewpoint, of course, ignores that long line of scientists, so prominent in the English-speaking tradition, who have always found their science and their religion mutually buttressing each other and who, whatever wounds they may have incidentally inflicted on a sound theology, have clearly had their scientific imagination stimulated by their religious concern. Those scientifically-minded readers whose interest in Newton is narrowly confined to his "natural" or "experimental philosophy," as contrasted with what may be called his broader and more speculative philosophy of nature, may rest content. Despite a large literature pointing to the theological antecedents of many of his more speculative concepts, and despite the strong influence on his thinking of the Cambridge circle stemming from Henry More, which included his teacher Isaac Barrow, it is hardly necessary to attribute Newton's more questionable assumptions—like the absolute motion, absolute space, and absolute time which awakened vigorous criticism in his own generation and have since been finally abandoned by physical theory in our own—to the intrusion of external theological ideas into his "experimental philosophy."

Doubtless Newton first learned about these absolutes from that attractive if none too consistent thinker Henry More, who began by bending the new Cartesian science to the service of his liberal or, as it was then called, "latitudinarian," theology, and then proceeded to demolish that science by reconstructing it to suit his own Platonic and irenic purposes. But Newton's absolutes scarcely required support in a modernistic theology, however deliciously bizarre. They are too firmly rooted in the mathematical procedures and principles of his science itself. Newton's physics needed extraneous theological and metaphysical "foundations" as little as our own, though it, like ours, surely had metaphysical implications; few natural scientists, it may be supposed, can be persuaded they are pursuing a physics without ontology. It seems quite clear that Newton's theological concepts were determined by his scientific ideas rather than vice versa. Theology, indeed, rarely distorts science; but science, in the hands of modernists like Newton, is always corrupting sound theology.

Modernistic or liberal theology, ever anxious to accommodate itself to the latest fashion in ideas, is at the present moment in some disrepute, since it is not intellectual acceptance but moral rejection of our world that seems of pressing religious concern to our most sensitive theologians and prophets. But our Western religions, which from the beginning have had to live with Greek thought, have never long been able to maintain a faith in serious conflict with the best available knowledge. Even shorter have been the intervals when the temporary expedient of assigning knowledge and faith to separate and mutually exclusive realms has remained successful. What Philo Judaeus, the Alexandrian doctors, and St. Augustine did with consummate skill, what Maimonides and St. Thomas did for a different science with no less skill, if less enduring success, Newton and his theological followers attempted once more in the eighteenth century. If their efforts in rational, or natural, religion were unfortunate rather than blessed with success, any theologian—whether he rejects the whole enterprise of natural theology or resolves to take his place in the long line of his predecessors who have likewise sought to adjust religious insight to modern knowledge—can learn much from a careful study of this particular episode in the history of religious thought. To read Newton will make him both more appreciative and more critical of Eddington, Jeans, and Whitehead.

If both the historian of science and the theologian, liberal or post-liberal, can profit from a reading of Newton, so too can the philosopher. Newton is one of those men, like Copernicus, Galileo, Darwin, Einstein, or Freud, whose ideas have made philosophers necessary and indeed have determined their problems. When such revolutionary thought is imposed on men whose thinking has been cast in another and more traditional mold, the impact is terrific; it starts reverberations in every direction. What do these ideas really mean? What are their further implications if we must actually live with them? How are they to be adjusted to what we have always believed and cherish deeply? Sometimes these insistently disturbing ideas are religious, as in the Hellenistic age and again at the time of the Reformation. Sometimes they are social, as in the age of the Sophists, in the flaming ideals of the French Revolution, and again in the crises of our own times. But during the modern period, though all reflection has been colored by the ideas born of the successive struggles for freedom and the recurrent waves of individualistic protest, and more recently by the mounting demand for security, it has been novel scientific conceptions, new notions of the world and of man himself, that have largely stimulated and determined the course of philosophic thinking. To understand the *history* of modern philosophy—that is, the problems it has confronted and wrestled with, the particular assumptions it has analyzed and those it has unconsciously accepted, the direction it has taken and the compromises at which it has arrived—we must turn to the major innovators in scientific thought. And the greatest of these, in historical importance and influence, has been Newton.

It is hardly too much to say that for a century after Newton nearly all philosophizing, especially in its more technical aspects, was an attempt to come to terms with, to analyze and criticize, and to extend Newton's thought. His philosophy of nature was an insistent fact to be reckoned with; this was the way things actually were.

> Nature and Nature's laws lay hid in night;
> God said, Let Newton be, and all was light.

His intellectual method was the voice of Science itself. To be sure, Newton spoke in two different keys; and according as men heard the one or the other the more clearly, they derived from him one or the other of the two major philosophies of science that came down from the Newtonian age to the nineteenth century. One aspect of his thinking was pushed by the British philosophers, from John Locke onward, and by common account reached its classic form in the naturalistic empiricism of David Hume. The other aspect was developed by the German physicists, like Euler, and the German analytical empiricists, and led to the imposing critical philosophy of Immanuel Kant, probably the most profound analysis of the assumptions of Newtonian mechanics. Any critical examination of Newton's own thought is sure to lead either to the position of Hume or that of Kant.

But even when the Romantic revolt and the storms of the French Revolution brought emotional, moral, and social problems to the center of attention and inspired more personal and non-scientific, if not actually anti-scientific philosophies, Newton's philosophy of nature still dominated men's thinking. It now became what men wanted to get away from or pass beyond. Newton may have been blind to the larger aspects of human experience, but he did describe Nature; and so that Newtonian Nature was supplemented with other more transcendent realms in which a free man might feel more at home. Whatever their further adventurings in the quest for Reality, the point of departure of the transcendental idealists remained Newton's Nature. Even the waves of evolutionary thought could not beat down the Newtonian granite shores. And our own revolution in physical theory, which has replaced biology as the source of disturbing intellectual problems, has for most men only served to re-establish Newton as the voice of Science. In 1952 as acute and intelligent a thinker as Joseph Wood Krutch could still think it worth while to argue seriously against the sufficiency of the Newtonian world and the Newtonian method, and of the "science" with which he continues to identify them.

The Newtonian philosophy of nature—as distinguished from Newton's "natural philosophy" or science—which for two centuries passed as the truth about our physical world, is now dead we are told, so dead that most of our own rapidly changing philosophies of nature discuss it without end and vie in pointing out just where Newton made the wrong assumptions and took the wrong turn. We cannot set forth what we believe today about nature or comprehend the significance of our beliefs without first examining in detail what we no longer believe. And the Newtonian philosophies of science—both of them—have now dropped out of our own experimental and pragmatic conceptions of scientific method, though the Humian strain is hardly neglected in our own logical empiricisms, and their most questionable assumptions are repetitions of those that led Newton and Locke to identical epistemological impasses. In any event, whatever our own conceptions of the nature of scientific verification and of the status and function of scientific theory, we can hardly even state them except against the background of a thorough grasp of the Newtonian theory.

Like the philosopher, the literary historian is confronted with the enormous repercussions of the ideas for which Newton was the symbol throughout the broad vistas of eighteenth-century poetry and literary expression. Concerned with what Newton came to stand for in the Age of Reason, he will be driven to ask what Newton actually said that managed to lend itself to so many curious ramifications. To him, what men made out of Newton's thought is of supreme significance. But if the process is fascinating, knowledge of the raw material is indispensable. Like the philosopher, the student of literature encounters the discrepancy between Newton the symbol and Newton's sophisticated, if less consistent, thought. How was Newton distorted, and what did he say to render himself liable to such mishandling?

The scientist, the theologian, the philosopher, and the literary historian would all do well to read what Newton himself said. The present selection has been made with a view to the interests of all of them. The scientist may complain that little of Newton's mathematical demonstration is included. If he has looked into the *Principia Mathematica*, he will realize that even with the notation modernized, as in Cajori, that book is one of the most difficult of all the scientific classics. Only a hardened reader of Great Books would venture upon it without guidance. Newton expressed himself so elliptically, with such lack of concern for the ordinary reader who could not fill in the missing steps for himself, that one is inclined to sympathize with the non-mathematically-minded John Locke, who, on the appearance of the *Principia*, was forced to ask his mathematical friends whether Newton's demonstrations could be relied upon, and when assured that they could, painfully tried to puzzle out the conclusions for himself. The editors have resolved to appeal to readers with the mathematical competence of Locke, who after all as a physician could claim respectability as a scientist. Readers who demand more can turn to Cajori, or, if they command Latin, to the earlier admirable editions of the Jesuits.

Newton was, as the phrase goes, a "seminal thinker." If we have here been more concerned to comment on the harvest than on the seed and its provenance, readers may exercise their own wits and their knowledge of the history of thought on the "sources" of his major ideas. Such an inquiry will take them far afield, and will lead to men like William of Ockham, Francesco Patrizzi, and Bernardino Telesio, as well as to sober scientists like Galileo. Newton being what he is, what started him off is of undying fascination. But still more important is what Newton himself started. It is the hope of the editors that those who properly appreciate what Newton started will be glad to learn just how he started it, by reading his own words—often a disconcerting process with a thinker whose originality and historical limitations have been so long buried under the easy disguise of Newton the symbol.

JOHN HERMAN RANDALL, JR.

SELECTIONS FROM NEWTON

Natural philosophy consists in discovering the frame and operations of nature, and reducing them, as far as may be, to general rules or laws—establishing these rules by observations and experiments, and thence deducing the causes and effects of things. . . . NEWTON

I. The Method of Natural Philosophy [a]

1. RULES OF REASONING IN PHILOSOPHY [b]

RULE I

We are to admit no more causes of natural things than such as are both true and sufficient to explain their appearances.

To this purpose the philosophers say that Nature does nothing in vain, and more is in vain when less will serve; for Nature is pleased with simplicity and affects not the pomp of superfluous causes.

RULE II

Therefore to the same natural effects we must, as far as possible, assign the same causes.

As to respiration in a man and in a beast, the descent of stones in Europe and in America, the light of our culinary fire and of the sun, the reflection of light in the earth and in the planets.

RULE III

The qualities of bodies, which admit neither intensification nor remission of degrees, and which are found to belong to all bodies within the reach of our experiments, are to be esteemed the universal qualities of all bodies whatsoever.

For since the qualities of bodies are only known to us by experiments, we are to hold for universal all such as universally agree with experiments, and such as are not liable to diminution can never be quite taken away. We are certainly not to relinquish the

---

a [Also on method, see the latter parts of Queries 28 and 29, in the Questions from the *Optics*, Part V.]

b [Rules of Reasoning in Philosophy. *Philosophiae Naturalis Principia Mathematica*, Bk. III, 1686.]

evidence of experiments for the sake of dreams and vain fictions of our own devising; nor are we to recede from the analogy of Nature, which is wont to be simple and always consonant to itself. We in no other way know the extension of bodies than by our senses, nor do these reach it in all bodies; but because we perceive extension in all that are sensible, therefore we ascribe it universally to all others also. That abundance of bodies are hard we learn by experience; and because the hardness of the whole arises from the hardness of the parts, we therefore justly infer the hardness of the undivided particles, not only of the bodies we feel, but of all others. That all bodies are impenetrable, we gather not from reason, but from sensation. The bodies which we handle we find impenetrable, and thence conclude impenetrability to be a universal property of all bodies whatsoever. That all bodies are movable and endowed with certain powers (which we call the inertia) of persevering in their motion, or in their rest, we only infer from the like properties observed in the bodies which we have seen. The extension, hardness, impenetrability, mobility, and inertia of the whole result from the extension, hardness, impenetrability, mobility, and inertia of the parts; and hence we conclude the least particles of all bodies to be also all extended, and hard and impenetrable, and movable, and endowed with their proper inertia. And this is the foundation of all philosophy. Moreover, that the divided but contiguous particles of bodies may be separated from one another is a matter of observation; and, in the particles that remain undivided, our minds are able to distinguish yet lesser parts, as is mathematically demonstrated. But whether the parts so distinguished and not yet divided may, by the powers of Nature, be actually divided and separated from one another we cannot certainly determine. Yet had we the proof of but one experiment that any undivided particle, in breaking a hard and solid body, suffered a division, we might by virtue of this rule conclude that the undivided as well as the divided particles may be divided and actually separated to infinity.

Lastly, if it universally appears, by experiments and astronomical observations, that all bodies about the earth gravitate toward the earth, and that in proportion to the quantity of matter which they severally contain; that the moon likewise, according to the quantity of its matter, gravitates toward the earth; that, on the other hand, our sea gravitates toward the moon; and all the planets one toward another; and the comets in like manner toward the sun: we must, in consequence of this rule, universally allow that all bodies whatsoever are endowed with a principle of mutual gravitation. For the argument from the appearances concludes with more force for the universal gravitation of all bodies than for their impenetrability, of which, among those in the celestial regions, we have no experiments nor any manner of observation. Not that I affirm gravity to be essential to bodies; by their *vis insita* I mean nothing but their inertia. This is immutable. Their gravity is diminished as they recede from the earth.

RULE IV

In experimental philosophy we are to look upon propositions inferred by general induction from phenomena as accurately or very nearly true, notwithstanding any contrary hypotheses that may be imagined, till such time as other phenomena occur by which they may either be made more accurate or liable to exceptions.

This rule we must follow, that the argument of induction may not be evaded by hypotheses.

2. ON HYPOTHESES

From a Letter to Oldenburg [c]

. . . For the best and safest method of philosophizing seems to be, first, to inquire diligently into the properties of things and to establish those properties by experiments, and to proceed later to hypotheses for the explanation of things themselves. For hypotheses ought to be applied only in the explanation of the properties

---

[c] [Newton's letter to Oldenburg. London, 1672. *Isaac Newtoni Opera quae exstant Omnia*, IV, p. 314.]

of things, and not made use of in determining them; except in so far as they may furnish experiments. And if anyone offers conjectures about the truth of things from the mere possibility of hypotheses, I do not see by what stipulation anything certain can be determined in any science; since one or another set of hypotheses may always be devised which will appear to supply new difficulties. Hence I judged that one should abstain from contemplating hypotheses, as from improper argumentation. . . .[1]

From Letters to Cotes [d]

I

I had yours of Feb. 18th, and the difficulty you mention which lies in these words, "since every attraction is mutual," is removed by considering that, as in geometry, the word 'hypothesis' is not taken in so large a sense as to include the axioms and postulates; so, in experimental philosophy, it is not to be taken in so large a sense as to include the first principles or axioms, which I call the laws of motion. These principles are deduced from phenomena and made general by induction, which is the highest evidence that a proposition can have in this philosophy. And the word 'hypothesis' is here used by me to signify only such a proposition as is not a phenomenon nor deduced from any phenomena, but assumed or supposed—without any experimental proof. Now the mutual and mutually equal attraction of bodies is a branch of the third law of motion, and how this branch is deduced from phenomena you may see at the end of the corollaries of the laws of motion. . . . If a body attracts another contiguous to it and is not mutually attracted by the other, the attracted body will drive the other before it, and both will go away together with an accelerated motion *in infinitum*, as it were, by a self-moving principle, contrary to the first law of motion, whereas there is no such phenomenon in all nature.

[d] [This and the next selection are from Newton's letters to Cotes. London, 1713. J. Edleston, *Correspondence of Sir Isaac Newton and Prof. Cotes*, pp. 154-56, 156-57.]

...And for preventing exceptions against the use of the word 'hypothesis,' I desire you to conclude the next paragraph in this manner: "For anything which is not deduced from phenomena ought to be called a hypothesis, and hypotheses of this kind, whether metaphysical or physical, whether of occult qualities or mechanical, have no place in experimental philosophy. In this philosophy, propositions are deduced from phenomena, and afterward made general by induction."...

II

On Saturday last I wrote to you, representing that experimental philosophy proceeds only upon phenomena and deduces general propositions from them only by induction. And such is the proof of mutual attraction. And the arguments for the impenetrability, mobility, and force of all bodies and for the laws of motion are no better. And he that in experimental philosophy would except against any of these must draw his objection from some experiment or phenomenon and not from a mere hypothesis, if the induction be of any force....[2]

3. THE EXPERIMENTAL METHOD

From a letter to Oldenburg [e]

... I cannot think it effectual for determining truth to examine the several ways by which phenomena may be explained, unless where there can be a perfect enumeration of all those ways. You know, the proper method for inquiring after the properties of things is to deduce them from experiments. And I told you that the theory which I propounded was evinced to me, not by inferring *'tis thus because not otherwise,* that is, not by deducing it only from a confutation of contrary suppositions, but by deriving it from experiments concluding positively and directly. The way therefore to examine it is by considering whether the experiments which I

[e] [From a letter to Oldenburg. July, 1672. *Opera Omnia* IV, pp. 320-21.]

propound do prove those parts of the theory to which they are applied, or by prosecuting other experiments which the theory may suggest for its examination.

To determine by these and such like queries seems the most proper and direct way to a conclusion. And therefore I could wish all objections were suspended from hypotheses or any other heads than these two: of showing the insufficiency of experiments to determine these queries, or prove any other parts of my theory, by assigning the flaws and defects in my conclusions drawn from them; or of producing other experiments which directly contradict me, if any such may seem to occur. For if the experiments which I urge be defective, it cannot be difficult to show the defects; but if valid, then by proving the theory, they must render all objections invalid.

II. Fundamental Principles of Natural Philosophy

1. NEWTON'S PREFACE TO THE FIRST EDITION OF THE *PRINCIPIA* [a]

Since the ancients (as we are told by Pappus [b]) esteemed the science of mechanics of greatest importance in the investigation of natural things, and the moderns, rejecting substantial forms and occult qualities, have endeavored to subject the phenomena of nature to the laws of mathematics, I have in this treatise cultivated mathematics as far as it relates to philosophy. The ancients considered mechanics in a twofold respect: as rational, which proceeds accurately by demonstration, and practical. To practical mechanics all the manual arts belong, from which mechanics took its name. But as artificers do not work with perfect accuracy, it comes to pass that mechanics is so distinguished from geometry that what is perfectly accurate is called geometrical; what is less so is called mechanical. However, the errors are not in the art, but in the artificers. He that works with less accuracy is an imperfect mechanic; and if any could work with perfect accuracy, he would be the most perfect mechanic of all; for the description of right lines and circles, upon which geometry is founded, belongs to mechanics. Geometry does not teach us to draw these lines, but requires them to be drawn; for it requires that the learner should first be taught to describe these accurately before he enters upon geometry, then it shows how by these operations problems may be solved. To describe right lines and circles are problems, but not

[a] [Written at Cambridge, Trinity College, May 8, 1686, the year of publication of the first edition.]

[b] [Pappus was the author of the *Synagoge* ("Collection"), the last great treatise of the Alexandrian mathematicians, end of the third century. The *Synagoge* was a guide to the study of Greek geometry. Many important Greek mathematical results have been preserved for later ages only through the work of Pappus. The *Synagoge* was written about 320 A.D.; Latin translation, 1589.]

9

geometrical problems. The solution of these problems is required from mechanics, and by geometry the use of them, when so solved, is shown; and it is the glory of geometry that from those few principles, brought from without, it is able to produce so many things. Therefore geometry is founded in mechanical practice and is nothing but that part of universal mechanics which accurately proposes and demonstrates the art of measuring. But since the manual arts are chiefly employed in the moving of bodies, it happens that geometry is commonly referred to their magnitude, and mechanics to their motion. In this sense rational mechanics will be the science of motions resulting from any forces whatsoever and of the forces required to produce any motions, accurately proposed and demonstrated. This part of mechanics, as far as it extended to the five powers which relate to manual arts, was cultivated by the ancients, who considered gravity (it not being a manual power) not otherwise than in moving weights by those powers. But I consider philosophy rather than arts, and write not concerning manual but natural powers, and consider chiefly those things which relate to gravity, levity, elastic force, the resistance of fluids, and the like forces, whether attractive or impulsive; and therefore I offer this work as the mathematical principles of philosophy, for the whole burden of philosophy seems to consist in this: from the phenomena of motions to investigate the forces of nature, and then from these forces to demonstrate the other phenomena; and to this end the general propositions in the First and Second Books are directed. In the Third Book I give an example of this in the explication of the System of the World; for by the propositions mathematically demonstrated in the former books, in the third I derive from the celestial phenomena the forces of gravity with which bodies tend to the sun and the several planets. Then from these forces, by other propositions which are also mathematical, I deduce the motions of the planets, the comets, the moon, and the sea. I wish we could derive the rest of the phenomena of Nature by the same kind of reasoning from mechanical principles, for I am induced by many reasons to suspect that they may all depend upon certain forces by which the particles of bodies, by some causes hitherto unknown, are either mutually impelled toward one another and cohere in regular figures, or are repelled and recede from one another. These forces being unknown, philosophers have hitherto attempted the search of Nature in vain; but I hope the principles here laid down will afford some light either to this or some truer method of philosophy.

In the publication of this work the most acute and universally learned Mr. Edmund Halley not only assisted me in correcting the errors of the press and preparing the geometrical figures, but it was through his solicitations that it came to be published; for when he had obtained of me my demonstrations of the figure of the celestial orbits, he continually pressed me to communicate the same to the Royal Society, who afterward, by their kind encouragement and entreaties, engaged me to think of publishing them. But after I had begun to consider the inequalities of the lunar motions, and had entered upon some other things relating to the laws and measures of gravity and other forces; and the figures that would be described by bodies attracted according to given laws; and the motion of several bodies moving among themselves; the motion of bodies in resisting mediums; the forces, densities, and motions of mediums; the orbits of the comets, and suchlike, I deferred that publication till I had made a search into those matters and could put forth the whole together. What relates to the lunar motions (being imperfect), I have put all together in the corollaries of Proposition LXVI, to avoid being obliged to propose and distinctly demonstrate the several things there contained in a method more prolix than the subject deserved and interrupt the series of the other propositions. Some things, found out after the rest, I chose to insert in places less suitable, rather than change the number of the propositions and the citations. I heartily beg that what I have here done may be read with forbearance and that my labors in a subject so difficult may be examined, not so much with the view to censure, as to remedy their defects.

Is. NEWTON

Cambridge, Trinity College, May 8, 1686

2. DEFINITIONS AND SCHOLIUM [e]

DEFINITION I

The quantity of matter is the measure of the same, arising from its density and bulk conjointly.

Thus air of a double density, in a double space, is quadruple in quantity; in a triple space, sextuple in quantity. The same thing is to be understood of snow and fine dust or powders that are condensed by compression or liquefaction, and of all bodies that are by any causes whatever differently condensed. I have no regard in this place to a medium, if any such there is, that freely pervades the interstices between the parts of bodies. It is this quantity that I mean hereafter everywhere under the name of 'body' or 'mass.' And the same is known by the weight of each body, for it is proportional to the weight, as I have found by experiments on pendulums, very accurately made, which shall be shown hereafter.

DEFINITION II

The quantity of motion is the measure of the same, arising from the velocity and quantity of matter conjointly.

The motion of the whole is the sum of the motions of all the parts; and therefore in a body double in quantity, with equal velocity, the motion is double; with twice the velocity, it is quadruple.

DEFINITION III

The vis insita, *or innate force of matter, is a power of resisting by which every body, as much as in it lies, continues in its present state, whether it be of rest or of moving uniformly forward in a right line.*

[e] [Definitions and Scholium to the Definitions, *Principia*, Bk. I.]

This force is always proportional to the body whose force it is and differs nothing from the inactivity of the mass, but in our manner of conceiving it. A body, from the inert nature of matter, is not without difficulty put out of its state of rest or motion. Upon which account, this *vis insita* may, by a most significant name, be called 'inertia' (*vis inertiae*) or 'force of inactivity.' But a body only exerts this force when another force, impressed upon it, endeavors to change its condition; and the exercise of this force may be considered as both resistance and impulse; it is resistance so far as the body, for maintaining its present state, opposes the force impressed; it is impulse so far as the body, by not easily giving way to the impressed force of another, endeavors to change the state of that other. Resistance is usually ascribed to bodies at rest, and impulse to those in motion; but motion and rest, as commonly conceived, are only relatively distinguished; nor are those bodies always truly at rest which commonly are taken to be so.

DEFINITION IV

An impressed force is an action exerted upon a body in order to change its state, either of rest or of uniform motion in a right line.

This force consists in the action only, and remains no longer in the body when the action is over. For a body maintains every new state it acquires by its inertia only. But impressed forces are of different origins, as from percussion, from pressure, from centripetal force.

DEFINITION V

A centripetal force is that by which bodies are drawn or impelled, or in any way tend toward a point as to a center.

Of this sort is gravity, by which bodies tend to the center of the earth; magnetism, by which iron tends to the loadstone; and that force, whatever it is, by which the planets are continually drawn aside from the rectilinear motions, which otherwise they would pursue, and made to revolve in curvilinear orbits. A stone, whirled about in a sling, endeavors to recede from the hand that turns it; and by that endeavor distends the sling, and that with so much the greater force as it is revolved with the greater velocity, and as soon as it is let go flies away. That force which opposes itself to this endeavor, and by which the sling continually draws back the stone toward the hand and retains it in its orbit, because it is directed to the hand as the center of the orbit, I call the centripetal force. And the same thing is to be understood of all bodies revolved in any orbits. They all endeavor to recede from the centers of their orbits; and were it not for the opposition of a contrary force which restrains them to and detains them in their orbits, which I therefore call centripetal, would fly off in right lines, with a uniform motion. A projectile, if it was not for the force of gravity, would not deviate toward the earth, but would go off from it in a right line, and that with a uniform motion if the resistance of the air was taken away. It is by its gravity that it is drawn aside continually from its rectilinear course and made to deviate toward the earth, more or less, according to the force of its gravity and the velocity of its motion. The less its gravity is or the quantity of its matter, or the greater the velocity with which it is projected, the less will it deviate from a rectilinear course and the farther it will go. If a leaden ball, projected from the top of a mountain by the force of gunpowder, with a given velocity and in a direction parallel to the horizon, is carried in a curved line to the distance of two miles before it falls to the ground; the same, if the resistance of the air were taken away, with a double or decuple velocity, would fly twice or ten times as far. And by increasing the velocity, we may at pleasure increase the distance to which it might be projected and diminish the curvature of the line which it might describe, till at last it should fall at the distance of 10, 30, or 90 degrees, or even might go quite round the whole earth before it falls; or lastly, so that it might never fall to the earth, but go forward into the celestial spaces, and proceed in its motion *in infinitum*. And after the same manner that a projectile, by the force of gravity, may be made to revolve in an orbit and go round the whole earth, the moon also, either by the force of gravity, if it is endued with gravity, or by any other force that impels it toward the earth, may be continually drawn aside toward the earth, out of the rectilinear way which by its innate force it would pursue, and would be made to revolve in the orbit which it now describes; nor could the moon without some such force be retained in its orbit. If this force was too small, it would not sufficiently turn the moon out of a rectilinear course; if it was too great, it would turn it too much and draw down the moon from its orbit toward the earth. It is necessary that the force be of a just quantity, and it belongs to the mathematicians to find the force that may serve exactly to retain a body in a given orbit with a given velocity; and, vice versa, to determine the curvilinear way into which a body projected from a given place, with a given velocity, may be made to deviate from its natural rectilinear way by means of a given force.

The quantity of any centripetal force may be considered as of three kinds: absolute, accelerative, and motive.

DEFINITION VI

The absolute quantity of a centripetal force is the measure of the same, proportional to the efficacy of the cause that propagates it from the center, through the spaces round about.

Thus the magnetic force is greater in one loadstone and less in another, according to their sizes and strength of intensity.

DEFINITION VII

The accelerative quantity of a centripetal force is the measure of the same, proportional to the velocity which it generates in a given time.

Thus the force of the same loadstone is greater at a less distance and less at a greater; also the force of gravity is greater in valleys, less on tops of exceeding high mountains, and yet less (as shall hereafter be shown) at greater distances from the body of the earth; but at equal distances, it is the same everywhere, because (taking away or allowing for the resistance of the air) it equally accelerates all falling bodies, whether heavy or light, great or small.

DEFINITION VIII

The motive quantity of a centripetal force is the measure of the same, proportional to the motion which it generates in a given time.

Thus the weight is greater in a greater body, less in a less body; and, in the same body, it is greater near to the earth and less at remoter distances. This sort of quantity is the centripetency or propension of the whole body towards the center, or, as I may say, its weight; and it is always known by the quantity of an equal and contrary force just sufficient to hinder the descent of the body.

These quantities of forces we may, for the sake of brevity, call by the names of 'motive,' 'accelerative,' and 'absolute forces'; and, for the sake of distinction, consider them with respect to the bodies that tend to the center, to the places of those bodies, and to the center of force toward which they tend; that is to say, I refer the motive force to the body as an endeavor and propensity of the whole toward a center, arising from the propensities of the several parts taken together; the accelerative force to the place of the body, as a certain power diffused from the center to all places around to move the bodies that are in them; and the absolute force to the center, as endued with some cause, without which those motive forces would not be propagated through the spaces round about; whether that cause be some central body (such as is the magnet in the center of the magnetic force or the earth in the center of the gravitating force) or anything else that does not yet appear. For I here design only to give a mathematical notion of those forces, without considering their physical causes and seats.

Wherefore the accelerative force will stand in the same relation to the motive as celerity does to motion. For the quantity of motion arises from the celerity multiplied by the quantity of matter, and the motive force arises from the accelerative force multiplied by the same quantity of matter. For the sum of the actions of the accelerative force, upon the several particles of the body, is the motive force of the whole. Hence it is that near the surface of the earth, where the accelerative gravity or force productive of gravity in all bodies is the same, the motive gravity or the weight is as the body; but if we should ascend to higher regions, where the accelerative gravity is less, the weight would be equally diminished and would always be as the product of the body by the accelerative gravity. So in those regions where the accelerative gravity is diminished into one half, the weight of a body two or three times less will be four or six times less.

I likewise call attractions and impulses, in the same sense, accelerative and motive; and use the words 'attraction,' 'impulse,' or 'propensity' of any sort toward a center, promiscuously and indifferently, one for another, considering those forces not physically but mathematically; wherefore the reader is not to imagine that by those words I anywhere take upon me to define the kind or the manner of any action, the causes or the physical reason thereof, or that I attribute forces, in a true and physical sense, to certain centers (which are only mathematical points) when at any time I happen to speak of centers as attracting or as endued with attractive powers.

SCHOLIUM [3]

Hitherto I have laid down the definitions of such words as are less known and explained the sense in which I would have them to be understood in the following discourse. I do not define time, space, place, and motion, as being well known to all. Only I must observe that the common people conceive those quantities under no other notions but from the relation they bear to sensible objects. And thence arise certain prejudices, for the removing of which it will be convenient to distinguish them into absolute and relative, true and apparent, mathematical and common.

1. Absolute, true, and mathematical time, of itself and from its own nature, flows equably without relation to anything external, and by another name is called 'duration'; relative, apparent, and common time is some sensible and external (whether accurate or unequable) measure of duration by the means of motion, which is commonly used instead of true time, such as an hour, a day, a month, a year.

2. Absolute space, in its own nature, without relation to any-

thing external, remains always similar and immovable. Relative space is some movable dimension or measure of the absolute spaces, which our senses determine by its position to bodies and which is commonly taken for immovable space; such is the dimension of a subterraneous, an aerial, or celestial space, determined by its position in respect of the earth. Absolute and relative space are the same in figure and magnitude, but they do not remain always numerically the same. For if the earth, for instance, moves, a space of our air, which relatively and in respect of the earth remains always the same, will at one time be one part of the absolute space into which the air passes; at another time it will be another part of the same, and so, absolutely understood, it will be continually changed.

3. Place is a part of space which a body takes up and is, according to the space, either absolute or relative. I say, a part of space; not the situation nor the external surface of the body. For the places of equal solids are always equal; but their surfaces, by reason of their dissimilar figures, are often unequal. Positions properly have no quantity; nor are they so much the places themselves as the properties of places. The motion of the whole is the same with the sum of the motions of the parts; that is, the translation of the whole, out of its place, is the same thing with the sum of the translations of the parts out of their places; and therefore the place of the whole is the same as the sum of the places of the parts, and for that reason it is internal and in the whole body.

4. Absolute motion is the translation of a body from one absolute place into another, and relative motion the translation from one relative place into another. Thus in a ship under sail the relative place of a body is that part of the ship which the body possesses, or that part of the cavity which the body fills and which therefore moves together with the ship, and relative rest is the continuance of the body in the same part of the ship or of its cavity. But real, absolute rest is the continuance of the body in the same part of that immovable space in which the ship itself, its cavity, and all that it contains is moved. Wherefore, if the earth is really at rest, the body, which relatively rests in the ship, will

really and absolutely move with the same velocity which the ship has on the earth. But if the earth also moves, the true and absolute motion of the body will arise, partly from the true motion of the earth in immovable space, partly from the relative motion of the ship on the earth; and if the body moves also relatively in the ship, its true motion will arise, partly from the true motion of the earth in immovable space and partly from the relative motions as well of the ship on the earth as of the body in the ship; and from these relative motions will arise the relative motion of the body on the earth. As if that part of the earth where the ship is was truly moved toward the east with a velocity of 10,010 parts, while the ship itself, with a fresh gale and full sails, is carried toward the west with a velocity expressed by 10 of those parts, but a sailor walks in the ship toward the east with 1 part of the said velocity; then the sailor will be moved truly in immovable space toward the east, with a velocity of 10,001 parts, and relatively on the earth toward the west, with a velocity of 9 of those parts.

Absolute time, in astronomy, is distinguished from relative by the equation or correction of the apparent time. For the natural days are truly unequal, though they are commonly considered as equal and used for a measure of time; astronomers correct this inequality that they may measure the celestial motions by a more accurate time. It may be that there is no such thing as an equable motion whereby time may be accurately measured. All motions may be accelerated and retarded, but the flowing of absolute time is not liable to any change. The duration or perseverance of the existence of things remains the same, whether the motions are swift or slow, or none at all; and therefore this duration ought to be distinguished from what are only sensible measures thereof and from which we deduce it, by means of the astronomical equation. The necessity of this equation, for determining the times of a phenomenon, is evinced as well from the experiments of the pendulum clock as by eclipses of the satellites of Jupiter.

As the order of the parts of time is immutable, so also is the order of the parts of space. Suppose those parts to be moved out of their places, and they will be moved (if the expression may be allowed) out of themselves. For times and spaces are, as it were, the places as well of themselves as of all other things. All things are placed in time as to order of succession and in space as to order of situation. It is from their essence or nature that they are places, and that the primary places of things should be movable is absurd. These are therefore the absolute places, and translations out of those places are the only absolute motions.

But because the parts of space cannot be seen or distinguished from one another by our senses, therefore in their stead we use sensible measures of them. For from the positions and distances of things from any body considered as immovable we define all places; and then, with respect to such places, we estimate all motions, considering bodies as transferred from some of those places into others. And so, instead of absolute places and motions, we use relative ones, and that without any inconvenience in common affairs; but in philosophical disquisitions, we ought to abstract from our senses and consider things themselves, distinct from what are only sensible measures of them. For it may be that there is no body really at rest to which the places and motions of others may be referred.

But we may distinguish rest and motion, absolute and relative, one from the other by their properties, causes, and effects. It is a property of rest that bodies really at rest do rest in respect to one another. And therefore, as it is possible that in the remote regions of the fixed stars, or perhaps far beyond them, there may be some body absolutely at rest, but impossible to know from the position of bodies to one another in our regions whether any of these do keep the same position to that remote body, it follows that absolute rest cannot be determined from the position of bodies in our regions.

It is a property of motion that the parts which retain given positions to their wholes do partake of the motions of those wholes. For all the parts of revolving bodies endeavor to recede from the axis of motion, and the impetus of bodies moving forward arises from the joint impetus of all the parts. Therefore, if surrounding bodies are moved, those that are relatively at rest within them will partake of their motion. Upon which account the true and absolute motion of a body cannot be determined by the translation of it from those which only seem to rest; for the external bodies ought not only to appear at rest, but to be really at rest. For otherwise all included bodies, besides their translation from near the surrounding ones, partake likewise of their true motions; and though that translation were not made, they would not be really at rest, but only seem to be so. For the surrounding bodies stand in the like relation to the surrounded as the exterior part of a whole does to the interior, or as the shell does to the kernel; but if the shell moves, the kernel will also move, as being part of the whole, without any removal from near the shell.

A property near akin to the preceding is this, that if a place is moved, whatever is placed therein moves along with it; and therefore a body which is moved from a place in motion partakes also of the motion of its place. Upon which account all motions, from places in motion, are no other than parts of entire and absolute motions; and every entire motion is composed of the motion of the body out of its first place and the motion of this place out of its place; and so on, until we come to some immovable place, as in the before-mentioned example of the sailor. Wherefore entire and absolute motions cannot be otherwise determined than by immovable places; and for that reason I did before refer those absolute motions to immovable places, but relative ones to movable places. Now no other places are immovable but those that, from infinity to infinity, do all retain the same given position one to another, and upon this account must ever remain unmoved and do thereby constitute immovable space.

The causes by which true and relative motions are distinguished, one from the other, are the forces impressed upon bodies to generate motion. True motion is neither generated nor altered but by some force impressed upon the body moved, but relative motion may be generated or altered without any force impressed upon the body. For it is sufficient only to impress some force on other bodies with which the former is compared that, by their giving way, that relation may be changed in which the relative rest or motion of this other body did consist. Again, true motion suffers always some change from any force impressed upon the moving body, but relative motion does not necessarily undergo any change by such forces. For if the same forces are likewise impressed on those other bodies with which the comparison is made, that the relative position may be preserved, then that condition will be preserved in which the relative motion consists. And therefore any relative motion may be changed when the true motion remains unaltered, and the relative may be preserved when the true suffers some change. Thus, true motion by no means consists in such relations.

The effects which distinguish absolute from relative motion are the forces of receding from the axis of circular motion. For there are no such forces in a circular motion purely relative, but in a true and absolute circular motion they are greater or less, according to the quanity of the motion. If a vessel, hung by a long cord, is so often turned about that the cord is strongly twisted, then filled with water and held at rest together with the water, thereupon by the sudden action of another force it is whirled about the contrary way, and while the cord is untwisting itself the vessel continues for some time in this motion, the surface of the water will at first be plain, as before the vessel began to move; but after that the vessel, by gradually communicating its motion to the water, will make it begin sensibly to revolve and recede by little and little from the middle, and ascend to the sides of the vessel, forming itself into a concave figure (as I have experienced); and the swifter the motion becomes, the higher will the water rise, till at last, performing its revolutions in the same times with the vessel, it becomes relatively at rest in it. This ascent of the water shows its endeavor to recede from the axis of its motion; and the true and absolute circular motion of the water, which is here directly contrary to the relative, becomes known and may be measured by this endeavor. At first, when the relative motion of the water in the vessel was greatest, it produced no endeavor to recede from the axis; the water showed no tendency to the circumference, nor any ascent toward the sides of the vessel, but remained of a plain surface, and therefore its true circular motion had not yet begun. But afterward, when the relative motion of the water had decreased, the ascent thereof toward the sides of the vessel proved its endeavor to recede from the axis; and this endeavor showed the real circular motion of the water continually increasing, till it had acquired its greatest quantity, when the water rested relatively in the vessel. And therefore this endeavor does not depend upon any translation of the water in respect of the ambient bodies; nor can true circular motion be defined by such translation. There is only one real circular motion of any one revolving body, corresponding to only one power of endeavoring to recede from its axis of motion, as its proper and adequate effect; but relative motions, in one and the same body, are innumerable, according to the various relations it bears to external bodies, and, like other relations, are altogether destitute of any real effect, any otherwise than they may perhaps partake of that one only true motion. And therefore in their system who suppose that our heavens, revolving below the sphere of the fixed stars, carry the planets along with them, the several parts of those heavens and the planets, which are indeed relatively at rest in their heavens, do yet really move. For they change their position one to another (which never happens to bodies truly at rest) and, being carried together with their heavens, partake of their motions and, as parts of revolving wholes, endeavor to recede from the axis of their motions.

Wherefore relative quantities are not the quantities themselves whose names they bear, but those sensible measures of them (either accurate or inaccurate) which are commonly used instead of the measured quantities themselves. And if the meaning of words is to be determined by their use, then by the names 'time,' 'space,' 'place,' and 'motion' their [sensible] measures are properly to be understood; and the expression will be unusual, and purely mathematical, if the measured quantities themselves are meant. On this account, those violate the accuracy of language, which ought to be kept precise, who interpret these words for the measured quantities. Nor do those less defile the purity of mathematical and philosophical truths who confound real quantities with their relations and sensible measures.

It is indeed a matter of great difficulty to discover and effectually to distinguish the true motions of particular bodies from the apparent, because the parts of that immovable space in which those motions are performed do by no means come under the observation of our senses. Yet the thing is not altogether desperate; for we have some arguments to guide us, partly from the apparent motions, which are the differences of the true motions; partly from the forces, which are the causes and effects of the true motions. For instance, if two globes, kept at a given distance one from the other by means of a cord that connects them, were revolved about their common center of gravity, we might, from the tension of the cord, discover the endeavor of the globes to recede from the axis of their motion, and from thence we might compute the quantity of their circular motions. And then if any equal forces should be impressed at once on the alternate faces of the globes to augment or diminish their circular motions, from the increase or decrease of the tension of the cord we might infer the increment or decrement of their motions, and thence would be found on what faces those forces ought to be impressed that the motions of the globes might be most augmented; that is, we might discover their hindmost faces, or those which, in the circular motion, do follow. But the faces which follow being known, and consequently the opposite ones that precede, we should likewise know the determination of their motions. And thus we might find both the quantity and the determination of this circular motion, even in an immense vacuum, where there was nothing external or sensible with which the globes could be compared. But now, if in that space some remote bodies were placed that kept always a given position one to another, as the fixed stars do in our regions, we could not indeed determine from the relative translation of the globes among those bodies whether the motion did belong to the globes or to the bodies. But if we observed the cord and found that its tension was that very tension which the motions of the globes required, we might conclude the motion to be in the globes and the bodies to be at rest; and then, lastly, from the translation of the globes among the bodies, we should find the determination of their motions. But how we are to obtain the true motions from their causes, effects, and apparent differences, and the converse, shall be explained more at large in the following treatise. For to this end it was that I composed it.

3. AXIOMS, OR LAWS OF MOTION [d]

LAW I

Every body continues in its state of rest or of uniform motion in a right line unless it is compelled to change that state by forces impressed upon it.

Projectiles continue in their motions, so far as they are not retarded by the resistance of the air or impelled downward by the force of gravity. A top, whose parts by their cohesion are continually drawn aside from rectilinear motions, does not cease its rotation otherwise than as it is retarded by the air. The greater bodies of the planets and comets, meeting with less resistance in freer spaces, preserve their motions both progressive and circular for a much longer time.

LAW II

The change of motion is proportional to the motive force impressed and is made in the direction of the right line in which that force is impressed.

If any force generates a motion, a double force will generate double the motion, a triple force triple the motion, whether that force be impressed altogether and at once or gradually and successively. And this motion (being always directed the same way with the generating force), if the body moved before, is added to or subtracted from the former motion, according as they directly conspire with or are directly contrary to each other; or obliquely joined, when they are oblique, so as to produce a new motion compounded from the determination of both.

[d] [*Principia,* Bk. I.]

LAW III

To every action there is always opposed an equal reaction; or, the mutual actions of two bodies upon each other are always equal and directed to contrary parts.

Whatever draws or presses another is as much drawn or pressed by that other. If you press a stone with your finger, the finger is also pressed by the stone. If a horse draws a stone tied to a rope, the horse (if I may so say) will be equally drawn back toward the stone; for the distended rope, by the same endeavor to relax or unbend itself, will draw the horse as much toward the stone as it does the stone toward the horse and will obstruct the progress of the one as much as it advances that of the other. If a body impinge upon another and by its force change the motion of the other, that body also (because of the equality of the mutual pressure) will undergo an equal change in its own motion, toward the contrary part. The changes made by these actions are equal, not in the velocities but in the motions of bodies; that is to say, if the bodies are not hindered by any other impediments. For, because the motions are equally changed, the changes of the velocities made toward contrary parts are inversely proportional to the bodies. This law takes place also in attractions, as will be proved in the next scholium.

4. THE MOTIONS OF BODIES [e]

COROLLARY I

A body, acted on by two forces simultaneously, will describe the diagonal of a parallelogram in the same time as it would describe the sides by those forces separately.

If a body in a given time, by the force M impressed apart in the place A, should with a uniform motion be carried from A to B,

[e] [*Principia,* Bk. I. Corollaries I, II, IV, V, VI, and Scholium to Corollary VI following the laws of motion.]

and by the force N impressed apart in the same place, should be carried from A to C, let the parallelogram ABCD be completed; and by both forces acting together, it will in the same time be carried in the diagonal from A to D. For since the force N acts in the direction of the line AC, parallel to BD, this force (by the Second Law) will not at all alter the velocity generated by the other force M, by which the body is carried toward the line BD. The body therefore will arrive at the line BD in the same time, whether the force N be

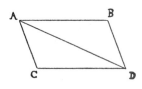

impressed or not; and therefore at the end of that time it will be found somewhere in the line BD. By the same argument, at the end of the same time it will be found somewhere in the line CD. Therefore it will be found in the point D, where both lines meet. But it will move in a right line from A to D, by Law I.

COROLLARY II

And hence is explained the composition of any one direct force AD out of any two oblique forces AC and CD; and, on the contrary, the resolution of any one direct force AD into two oblique forces AC and CD: which composition and resolution are abundantly confirmed from mechanics.

COROLLARY IV

The common center of gravity of two or more bodies does not alter its state of motion or rest by the actions of the bodies among themselves, and therefore the common center of gravity of all bodies acting upon each other (excluding external actions and impediments) is either at rest or moves uniformly in a right line.

For if two points proceed with a uniform motion in right lines, and their distance be divided in a given ratio, the dividing point will be either at rest or proceed uniformly in a right line. This is demonstrated hereafter in Lemma XXIII and Corollary, when the points are moved in the same plane; and, by a like way of arguing, it may be demonstrated when the points are not moved in the same plane. Therefore if any number of bodies move uniformly in right lines, the common center of gravity of any two of them is either at rest or proceeds uniformly in a right line, because the line which connects the centers of those two bodies so moving is divided at that common center in a given ratio. In like manner the common center of those two and that of a third body will be either at rest or moving uniformly in a right line, because at that center the distance between the common center of the two bodies and the center of this last is divided in a given ratio. In like manner the common center of these three and of a fourth body is either at rest or moves uniformly in a right line, because the distance between the common center of the three bodies and the center of the fourth is there also divided in a given ratio, and so on *in infinitum*. Therefore, in a system of bodies where there is neither any mutual action among themselves nor any foreign force impressed upon them from without, and which consequently move uniformly in right lines, the common center of gravity of them all is either at rest or moves uniformly forward in a right line.

Moreover, in a system of two bodies acting upon each other, since the distances between their centers and the common center of gravity of both are reciprocally as the bodies, the relative motions of those bodies, whether of approaching to or of receding from that center, will be equal among themselves. Therefore, since the changes which happen to motions are equal and directed to contrary parts, the common center of those bodies, by their mutual action between themselves, is neither accelerated nor retarded, nor suffers any change as to its state of motion or rest. But in a system of several bodies, because the common center of gravity of any two acting upon each other suffers no change in its state by that action, and much less the common center of gravity of the others with which that action does not intervene; but the distance between those two centers is divided by the common center of gravity of all the bodies into parts inversely proportional to the total sums of those bodies whose centers they are; and therefore while those two centers retain their state of motion or rest, the common center of all does also retain its state: it is manifest that the common center of all never suffers any change in the state of its motion or rest from the actions of any two bodies between themselves. But in such a system all the actions of the bodies among themselves either happen between two bodies or are composed of actions interchanged between some two bodies; and therefore they do never produce any alteration in the common center of all as to its state of motion or rest. Wherefore, since that center, when the bodies do not act one upon another, either is at rest or moves uniformly forward in some right line, it will, notwithstanding the mutual actions of the bodies among themselves, always continue in its state, either of rest or of proceeding uniformly in a right line, unless it is forced out of this state by the action of some power impressed from without upon the whole system. And therefore the same law takes place, in a system consisting of many bodies as in one single body, with regard to their persevering in their state of motion or of rest. For the progressive motion, whether of one single body or of a whole system of bodies, is always to be estimated from the motion of the center of gravity.

COROLLARY V

The motions of bodies included in a given space are the same among themselves, whether that space is at rest or moves uniformly forward in a right line without any circular motion.

For the differences of the motions tending toward the same parts and the sums of those that tend toward contrary parts are, at first (by supposition), in both cases the same; and it is from those sums and differences that the collisions and impulses do arise with which the bodies impinge one upon another. Wherefore (by Law II), the effects of those collisions will be equal in both cases, and therefore the mutual motions of the bodies among themselves in the one case will remain equal to the motions of the bodies among themselves in the other. A clear proof of this we have from the experiment of a ship, where all motions happen after the same manner, whether the ship is at rest or is carried uniformly forward in a right line.

COROLLARY VI

If bodies, moved in any manner among themselves, are urged in the direction of parallel lines by equal accelerative forces, they will all continue to move among themselves, after the same manner as if they had not been urged by those forces.

For these forces, acting equally (with respect to the quantities of the bodies to be moved) and in the direction of parallel lines, will (by Law II) move all the bodies equally (as to velocity), and therefore will never produce any change in the positions or motions of the bodies among themselves.

SCHOLIUM

Hitherto I have laid down such principles as have been received by mathematicians and are confirmed by abundance of experiments. By the first two laws and the first two corollaries, Galileo discovered that the descent of bodies varied as the square of the time (*in duplicata ratione temporis*) and that the motion of projectiles was in the curve of a parabola, experience agreeing with both, unless so far as these motions are a little retarded by the resistance of the air. When a body is falling, the uniform force of its gravity, acting equally, impresses in equal intervals of time equal forces upon that body, and therefore generates equal velocities; and, in the whole time impresses a whole force and generates a whole velocity proportional to the time. And the spaces described in proportional times are as the product of the velocities and the times, that is, as the squares of the times. And when a body is thrown upward, its uniform gravity impresses forces and reduces velocities proportional to the times; and the times of ascending to the greatest heights are as the velocities to be taken away, and those heights are as the product of the velocities and the times or as the squares of the velocities. And if a body be projected in any direction, the motion arising from its projection is compounded with the motion arising from its gravity. Thus, if the body A, by its motion of projection alone, could describe in a given time the right line AB, and with its motion of falling alone could describe in the same time the altitude AC, complete the parallelogram ABCD; and the body by that compounded motion will at the end of the time be found in the place D, and the curved line AED which that body describes will be a parabola, to which the right line AB will be a tangent at A and whose ordinate BD will be as the square of the line AB. On the same laws and corollaries depend those things which have been demonstrated concerning the times of the vibration of pendulums and are confirmed by the daily experiments of pendulum clocks. By the same, together with Law III, Sir Christopher Wren, Dr. Wallis, and Mr. Huygens, the greatest geometers of our times, did sev-

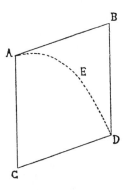

erally determine the rules of the impact and reflection of hard bodies and about the same time communicated their discoveries to the Royal Society, exactly agreeing among themselves as to those rules. Dr. Wallis, indeed, was somewhat earlier in the publication; then followed Sir Christopher Wren and lastly, Mr. Huygens. But Sir Christopher Wren confirmed the truth of the thing before the Royal Society by the experiments on pendulums, which M. Mariotte soon after thought fit to explain in a treatise entirely upon that subject. But to bring this experiment to an accurate agreement with the theory, we are to have due regard as well to the resistance of the air as to the elastic force of the concurring bodies. Let the spherical

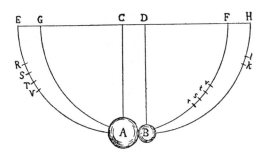

bodies A, B be suspended by the parallel and equal strings AC, BD from the centers C, D. About these centers, with those lengths as radii, describe the semicircles EAF, GBH, bisected respectively by the radii CA, DB. Bring the body A to any point R of the arc EAF and (withdrawing the body B) let it go from thence, and after one oscillation suppose it to return to the point V; then RV will be the retardation arising from the resistance of the air. Of this RV let ST be a fourth part, situated in the middle, namely, so that

$$RS = TV, \text{ and } RS : ST = 3 : 2,$$

then will ST represent very nearly the retardation during the descent from S to A. Restore the body B to its place; and, supposing the body A to be let fall from the point S, the velocity thereof in the place of reflection A, without sensible error, will be the same as if it had descended *in vacuo* from the point T. Upon which account this velocity may be represented by the chord of the arc TA. For it is a proposition well known to geometers that the velocity of a pendulous body in the lowest point is as the chord of the arc which it has described in its descent. After reflection, suppose the body A comes to the place *s* and the body B to the place *k*. Withdraw the body B and find the place *v*, from which if the body A, being let go, should after one oscillation return to the place *r*, *st* may be a fourth part of *rv*, so placed in the middle thereof as to leave *rs* equal to *tv*, and let the chord of the arc *t*A represent the velocity which the body A had in the place A immediately after reflection. For *t* will be the true and correct place to which the body A should have ascended if the resistance of the air had been taken off. In the same way we are to correct the place *k* to which the body B ascends by finding the place *l* to which it should have ascended *in vacuo*. And thus everything may be subjected to experiment in the same manner as if we were really placed in *vacuo*. These things being done, we are to take the product (if I may say so) of the body A by the chord of the arc TA (which represents its velocity), that we may have its motion in the place A immediately before reflection; and then by the chord of the arc *t*A, that we may have its motion in the place A immediately after reflection. And so we are to take the product of the body B by the chord of the arc Bl, that we may have the motion of the same immediately after reflection. And, in like manner, when two bodies are let go together from different places, we are to find the motion of each, as well before as after reflection; and then we may compare the motions between themselves and collect the effects of the reflection. Thus trying the thing with pendulums of 10 feet, in unequal as well as equal bodies, and making the bodies to concur after a descent through large spaces, as of 8, 12, or 16 feet, I found always, without an error of 3 inches, that when the bodies concurred together directly equal changes toward the contrary parts were produced in their motions and, of consequence, that the action and reaction were always equal. As if the body A impinged upon the body B at rest with 9 parts of motion and, losing 7, proceeded after reflection with 2, the body B was carried backward with those 7 parts. If the bodies concurred with contrary motions, A with 12 parts of motion and B with 6, then if A receded with 2, B receded with 8, namely, with a deduction of 14 parts of motion on each side. For from the motion of A, subtracting 12 parts, nothing will remain; but subtracting 2 parts more, a motion will be generated of 2 parts toward the contrary way; and so, from the motion of the body B of 6 parts, subtracting 14 parts, a motion is generated of 8 parts toward the contrary way. But if the bodies were made both to move toward the same way, A, the swifter, with 14 parts of motion, B, the slower, with 5, and after reflection A went on with 5, B likewise went on with 14 parts, 9 parts being transferred from A to B. And so in other cases. By the meeting and collision of bodies, the quantity of motion, obtained from the sum of the motions directed toward the same way, or from the difference of those that were directed toward contrary ways, was never changed. For the error of an inch or two in measures may be easily ascribed to the difficulty of executing everything with accuracy. It was not easy to let go the two pendulums so exactly together that the bodies should impinge one upon the other in the lowermost place AB; nor to mark the places s and k, to which the bodies ascended after impact. Nay, and some errors, too, might have happened from the unequal density of the parts of the pendulous bodies themselves and from the irregularity of the texture proceeding from other causes.

But to prevent an objection that may perhaps be alleged against the rule for the proof of which this experiment was made, as if this rule did suppose that the bodies were either absolutely hard or at least perfectly elastic (whereas no such bodies are to be found in Nature), I must add that the experiments we have been describing, by no means depending upon that quality of hardness, do succeed as well in soft as in hard bodies. For if the rule is to be tried in bodies not perfectly hard, we are only to diminish the reflection in such a certain proportion as the quantity of the elastic force requires. By the theory of Wren and Huygens, bodies absolutely hard return one from another with the same velocity with which they meet. But this may be affirmed with more certainty of bodies perfectly elastic. In bodies imperfectly elastic, the velocity of the return is to be diminished together with the elastic force, because that force (except when the parts of bodies are bruised by their impact or suffer some such extension as happens under the strokes of a hammer) is (as far as I can perceive) certain and determined, and makes the bodies to return one from the other with a relative velocity which is in a given ratio to that relative velocity with which they met. This I tried in balls of wool, made up tightly, and strongly compressed. For, first, by letting go the pendulous bodies and measuring their reflection, I determined the quantity of their elastic force; and then, according to this force, estimated the reflections that ought to happen in other cases of impact. And with this computation other experiments made afterward did accordingly agree, the balls always receding one from the other with a relative velocity which was to the relative velocity with which they met as about 5 to 9. Balls of steel returned with almost the same velocity, those of cork with a velocity something less; but in balls of glass the proportion was as about 15 to 16. And thus the Third Law, so far as it regards percussions and reflections, is proved by a theory exactly agreeing with experience.

In attractions, I briefly demonstrate the thing after this manner. Suppose an obstacle is interposed to hinder the meeting of any two bodies A, B, attracting one the other; then if either body, as

A, is more attracted toward the other body B than that other body B is toward the first body A, the obstacle will be more strongly urged by the pressure of the body A than by the pressure of the body B, and therefore will not remain in equilibrium; but the stronger pressure will prevail and will make the system of the two bodies, together with the obstacle, to move directly toward the parts on which B lies, and in free spaces to go forward *in infinitum* with a motion continually accelerated, which is absurd and contrary to the First Law. For, by the First Law, the system ought to continue in its state of rest or of moving uniformly forward in a right line; and therefore the bodies must equally press the obstacle and be equally attracted one by the other. I made the experiment on the loadstone and iron. If these, placed apart in proper vessels, are made to float by one another in standing water, neither of them will propel the other; but, by being equally attracted, they will sustain each other's pressure, and rest at last in an equilibrium.

So the gravitation between the earth and its parts is mutual. Let the earth FI be cut by any plane EG into two parts EGF and EGI, and their weights one toward the other will be mutually equal. For if by another plane HK, parallel to the former EG, the greater part EGI is cut into two parts EGKH and HKI, whereof HKI is equal to the part EFG, first cut off, it is evident that the middle part EGKH will have no propension by its proper weight toward either side, but will hang, as it were, and

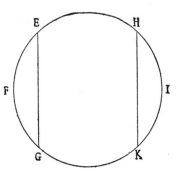

rest in an equilibrium between both. But the one extreme part HKI will with its whole weight bear upon and press the middle part toward the other extreme part EGF; and therefore the force with which EGI, the sum of the parts HKI and EGKH, tends toward the third part EGF, is equal to the weight of the part HKI, that is, to the weight of the third part EGF. And therefore the weights of the two parts EGI and EGF, one toward the other, are equal, as I was to prove. And indeed if those weights were not equal, the whole earth floating in the nonresisting ether would give way to the greater weight and, retiring from it, would be carried off *in infinitum*.

And as those bodies are equipollent in the impact and reflection whose velocities are inversely as their innate forces, so in the use of mechanic instruments those agents are equipollent and mutually sustain each the contrary pressure of the other whose velocities, estimated according to the determination of the forces, are inversely as the forces.

So those weights are of equal force to move the arms of a balance which, during the play of the balance, are inversely as their velocities upward and downward; that is, if the ascent or descent is direct, those weights are of equal force which are inversely as the distances of the points at which they are suspended from the axis of the balance; but if they are turned aside by the interposition of oblique planes or other obstacles and made to ascend or descend obliquely, those bodies will be equipollent which are inversely as the heights of their ascent and descent, taken according to the perpendicular, and that on account of the determination of gravity downward.

And in like manner in the pulley, or in a combination of pulleys, the force of a hand drawing the rope directly, which is to the weight, whether ascending directly or obliquely, as the velocity of the perpendicular ascent of the weight to the velocity of the hand that draws the rope, will sustain the weight.

In clocks and suchlike instruments, made up from a combination of wheels, the contrary forces that promote and impede the motion of the wheels, if they are inversely as the velocities of the parts of the wheel on which they are impressed, will mutually sustain each other.

The force of the screw to press a body is to the force of the hand that turns the handles by which it is moved as the circular velocity of the handle, in that part where it is impelled by the hand, is to the progressive velocity of the screw toward the pressed body.

The forces by which the wedge presses or drives the two parts of the wood it cleaves are to the force of the mallet upon the wedge as the progress of the wedge in the direction of the force impressed upon it by the mallet is to the velocity with which the parts of the wood yield to the wedge, in the direction of lines perpendicular to the sides of the wedge. And the like account is to be given of all machines.

The power and use of machines consist only in this, that by diminishing the velocity we may augment the force, and the contrary; from whence, in all sorts of proper machines, we have the solution of this problem: *To move a given weight with a given power*, or with a given force to overcome any other given resistance. For if machines are so contrived that the velocities of the agent and resistant are inversely as their forces, the agent will just sustain the resistant, but with a greater disparity of velocity will overcome it. So that if the disparity of velocities is so great as to overcome all that resistance which commonly arises either from the friction of contiguous bodies as they slide by one another, or from the cohesion of continuous bodies that are to be separated, or from the weights of bodies to be raised, the excess of the force remaining, after all those resistances are overcome, will produce an acceleration of motion proportional thereto, as well in the parts of the machine as in the resisting body. But to treat of mechanics is not my present business. I was aiming only to show by those examples the great extent and certainty of the Third Law of Motion. For if we estimate the action of the agent from the product of its force and velocity, and likewise the reaction of the impediment from the product of the velocities of its several parts, and the forces of resistance arising from the friction, cohesion, weight, and acceleration of those parts, the action and reaction in the use of all sorts of machines will be found always equal to one another. And so far as the action is propagated by the intervening instruments, and at last impressed upon the resisting body, the ultimate action will be always contrary to the reaction.

PROPOSITION LXIX, THEOREM XXIX [f]

In a system of several bodies A, B, C, D, etc., if any one of those bodies, as A, attract all the rest, B, C, D, etc., with accelerative

[f] [Proposition LXIX, Theorem XXIX, and Corollaries I, II, III; and Scholium following this Proposition, *Principia*, Bk. I.]

forces that are inversely as the squares of the distances from the attracting body, and another body, as B, attracts also the rest, A, C, D, etc., with forces that are inversely as the squares of the distances from the attracting body, the absolute forces of the attracting bodies A and B will be to each other as those very bodies A and B to which those forces belong.

For the accelerative attractions of all the bodies B, C, D toward A are by the supposition equal to each other at equal distances; and in like manner the accelerative attractions of all the bodies toward B are also equal to each other at equal distances. But the absolute attractive force of the body A is to the absolute attractive force of the body B as the accelerative attraction of all the bodies toward A is to the accelerative attraction of all the bodies toward B at equal distances; and so is also the accelerative attraction of the body B toward A to the accelerative attraction of the body A toward B. But the accelerative attraction of the body B toward A is to the accelerative attraction of the body A toward B as the mass of the body A is to the mass of the body B; because the motive forces which (by the Second, Seventh, and Eighth Definitions) are as the accelerative forces and the bodies attracted conjointly are here equal to one another by the Third Law. Therefore the absolute attractive force of the body A is to the absolute attractive force of the body B as the mass of the body A is to the mass of the body B. Q.E.D.

COROLLARY I

Therefore if each of the bodies of the system A, B, C, D, etc. does singly attract all the rest with accelerative forces that are inversely as the squares of the distances from the attracting body, the absolute forces of all those bodies will be to each other as the bodies themselves.

COROLLARY II

By a like reasoning, if each of the bodies of the system A, B, C, D, etc. does singly attract all the rest with accelerative forces, which are either inversely or directly in the ratio of any power whatever of the distances from the attracting body, or which are defined by the distances from each of the attracting bodies according to any common law, it is plain that the absolute forces of those bodies are as the bodies themselves.

COROLLARY III

In a system of bodies whose forces decrease as the square of the distances, if the lesser revolve about one very great one in ellipses, having their common focus in the center of that great body, and of a figure exceedingly accurate, and moreover by radii drawn to that great body describe areas proportional to the times exactly, the absolute forces of those bodies to each other will be either accurately or very nearly in the ratio of the bodies. And so conversely. . . .

SCHOLIUM

These Propositions naturally lead us to the analogy there is between centripetal forces and the central bodies to which those forces are usually directed; for it is reasonable to suppose that forces which are directed to bodies should depend upon the nature and quantity of those bodies, as we see they do in magnetical experiments. And when such cases occur, we are to compute the attractions of the bodies by assigning to each of their particles its proper force, and then finding the sum of them all. I here use the word 'attraction' in general for any endeavor whatever made by bodies to approach to each other, whether that endeavor arise from the action of the bodies themselves, as tending to each other or agitating each other by spirits emitted; or whether it arises from the action of the ether or of the air, or of any medium whatever, whether corporeal or incorporeal, in any manner impelling bodies placed therein toward each other. In the same general sense I use the word 'impulse,' not defining in this treatise the species or physical qualities of forces, but investigating the quantities and mathematical proportions of them, as I observed before in the definitions. In mathematics we are to investigate the quantities of forces with their proportions consequent upon any conditions supposed; then, when we enter upon physics, we compare those proportions with the phenomena of Nature, that we may know what conditions of those forces answer to the several kinds of attractive bodies. And this preparation being made, we argue more safely concerning the physical species, causes, and proportions of the forces. . . .

III. God and Natural Philosophy [a]

1. GENERAL SCHOLIUM [b]

The hypothesis of vortices is pressed with many difficulties. That every planet by a radius drawn to the sun may describe areas proportional to the times of description, the periodic times of the several parts of the vortices should observe the square of their distances from the sun; but that the periodic times of the planets may obtain the $3/2$th power of their distances from the sun, the periodic times of the parts of the vortex ought to be as the $3/2$th power of their distances. That the smaller vortices may maintain their lesser revolutions about Saturn, Jupiter, and other planets, and swim quietly and undisturbed in the greater vortex of the sun, the periodic times of the parts of the sun's vortex should be equal; but the rotation of the sun and planets about their axes, which ought to correspond with the motions of their vortices, recede far from all these proportions. The motions of the comets are exceedingly regular, are governed by the same laws with the motions of the planets, and can by no means be accounted for by the hypothesis of vortices; for comets are carried with very eccentric motions through all parts of the heavens indifferently, with a freedom that is incompatible with the notion of a vortex.

Bodies projected in our air suffer no resistance but from the air. Withdraw the air, as is done in Mr. Boyle's vacuum, and the resistance ceases; for in this void a bit of fine down and a piece of solid gold descend with equal velocity. And the same argument must apply to the celestial spaces above the earth's atmosphere; in these spaces, where there is no air to resist their motions, all bodies will move with the greatest freedom; and the planets and comets will constantly pursue their revolutions in orbits given in kind and position, according to the laws above explained; but though these

a [See also the Questions from the *Optics,* Part V, 28 and 31.]
b [*Principia,* Bk. III. See also Note 4, p. 183.]

bodies may, indeed, continue in their orbits by the mere laws of gravity, yet they could by no means have at first derived the regular position of the orbits themselves from those laws.

The six primary planets are revolved about the sun in circles concentric with the sun, and with motions directed toward the same parts and almost in the same plane. Ten moons are revolved about the earth, Jupiter, and Saturn, in circles concentric with them, with the same direction of motion, and nearly in the planes of the orbits of those planets; but it is not to be conceived that mere mechanical causes could give birth to so many regular motions, since the comets range over all parts of the heavens in very eccentric orbits; for by that kind of motion they pass easily through the orbs of the planets, and with great rapidity; and in their aphelions, where they move the slowest and are detained the longest, they recede to the greatest distances from each other, and hence suffer the least disturbance from their mutual attractions. This most beautiful system of the sun, planets, and comets could only proceed from the counsel and dominion of an intelligent and powerful Being. And if the fixed stars are the centers of other like systems, these, being formed by the like wise counsel, must be all subject to the dominion of One, especially since the light of the fixed stars is of the same nature with the light of the sun and from every system light passes into all the other systems; and lest the systems of the fixed stars should, by their gravity, fall on each other, he hath placed those systems at immense distances from one another.

This Being governs all things, not as the soul of the world, but as Lord over all; and on account of his dominion he is wont to be called "Lord God" παντοκράτωρ, or "Universal Ruler"; for 'God' is a relative word and has a respect to servants; and Deity is the dominion of God, not over his own body, as those imagine who fancy God to be the soul of the world, but over servants. The Supreme God is a Being eternal, infinite, absolutely perfect, but a being, however perfect, without dominion, cannot be said to be "Lord God"; for we say "my God," "your God," "the God of Israel," "the God of Gods," and "Lord of Lords," but we do not say "my Eternal," "your Eternal," "the Eternal of Israel," "the Eternal of Gods"; we do not say "my Infinite," or "my Perfect": these are titles which have no respect to servants. The word 'God' [c] usually signifies 'Lord,' but every lord is not a God. It is the dominion of a spiritual being which constitutes a God; a true, supreme, or imaginary dominion makes a true, supreme, or imaginary God. And from his true dominion it follows that the true God is a living, intelligent, and powerful Being; and, from his other perfections, that he is supreme or most perfect. He is eternal and infinite, omnipotent and omniscient; that is, his duration reaches from eternity to eternity; his presence from infinity to infinity; he governs all things and knows all things that are or can be done. He is not eternity and infinity, but eternal and infinite; he is not duration or space, but he endures and is present. He endures forever and is everywhere present; and, by existing always and everywhere, he constitutes duration and space. Since every particle of space is *always*, and every indivisible moment of duration is *everywhere*, certainly the Maker and Lord of all things cannot be *never* and *nowhere*. Every soul that has perception is, though in different times and in different organs of sense and motion, still the same indivisible person. There are given successive parts in duration, coexistent parts in space, but neither the one nor the other in the person of a man or his thinking principle; and much less can they be found in the thinking substance of God. Every man, so far as he is a thing that has perception, is one and the same man during his whole life, in all and each of his organs of sense. God is the same God, always and everywhere. He is omnipresent not *virtually* only but also *substantially;* for virtue cannot subsist without substance. In him [d] are all things contained and moved, yet neither

[c] Dr. Pocock derives the Latin word '*Deus*' from the Arabic '*du*' (in the oblique case '*di*'), which signifies 'Lord.' And in this sense princes are called "gods," Psalm lxxxii. ver. 6; and John x. ver. 35. And Moses is called a "god" to his brother Aaron, and a "god" to Pharaoh (Exodus iv. ver. 16; and vii. ver. 1). And in the same sense the souls of dead princes were formerly, by the Heathens, called "gods," but falsely, because of their want of dominion.

[d] This was the opinion of the Ancients. So Pythagoras, in *Cicer. de Nat. Deor.* lib. i. Thales, Anaxagoras, Virgil, *Georg.* lib. iv. ver. 220; and *Aeneid,* lib. vi. ver. 721. *Philo Allegor,* at the beginning of lib. i. Aratus, in his *Phaenom,* at the beginning. So also the sacred writers: as St. Paul, Acts xvii. ver. 27, 28. St. John's Gosp. chap. xiv. ver. 2. Moses, in Deuteronomy iv. ver. 39; and x. ver. 14. David, Psalm cxxxix, ver. 7, 8, 9. Solomon, 1 Kings viii. ver. 27. Job

affects the other; God suffers nothing from the motion of bodies, bodies find no resistance from the omnipresence of God. It is allowed by all that the Supreme God exists necessarily, and by the same necessity he exists *always* and *everywhere*. Whence also he is all similar, all eye, all ear, all brain, all arm, all power to perceive, to understand, and to act; but in a manner not at all human, in a manner not at all corporeal, in a manner utterly unknown to us. As a blind man has no idea of colors, so have we no idea of the manner by which the all-wise God perceives and understands all things. He is utterly void of all body and bodily figure, and can therefore neither be seen nor heard nor touched; nor ought he to be worshiped under the representation of any corporeal thing. We have ideas of his attributes, but what the real substance of anything is we know not. In bodies we see only their figures and colors, we hear only the sounds, we touch only their outward surfaces, we smell only the smells and taste the savors, but their inward substances are not to be known either by our senses or by any reflex act of our minds; much less, then, have we any idea of the substance of God. We know him only by his most wise and excellent contrivances of things and final causes; we admire him for his perfections, but we reverence and adore him on account of his dominion, for we adore him as his servants; and a god without dominion, providence, and final causes is nothing else but Fate and Nature. Blind metaphysical necessity, which is certainly the same always and everywhere, could produce no variety of things. All that diversity of natural things which we find suited to different times and places could arise from nothing but the ideas and will of a Being necessarily existing. But, by way of allegory, God is said to see, to speak, to laugh, to love, to hate, to desire, to give, to receive, to rejoice, to be angry, to fight, to frame, to work, to build; for all our notions of God are taken from the ways of mankind by a certain similitude, which, though not perfect, has some likeness, however. And thus much concerning God, to discourse of whom

---

xxii. ver. 12, 13, 14. Jeremiah, xxiii. ver. 23, 24. The Idolaters supposed the sun, moon, and stars, the souls of men, and other parts of the world to be parts of the Supreme God, and therefore to be worshiped; but erroneously.

from the appearances of things does certainly belong to natural philosophy.

Hitherto we have explained the phenomena of the heavens and of our sea by the power of gravity, but have not yet assigned the cause of this power. This is certain, that it must proceed from a cause that penetrates to the very centers of the sun and planets, without suffering the least diminution of its force; that operates not according to the quanity of the surfaces of the particles upon which it acts (as mechanical causes used to do), but according to the quantity of the solid matter which they contain, and propagates its virtue on all sides to immense distances, decreasing always as the inverse square of the distances. Gravitation toward the sun is made up out of the gravitations toward the several particles of which the body of the sun is composed, and in receding from the sun decreases accurately as the inverse square of the distances as far as the orbit of Saturn, as evidently appears from the quiescence of the aphelion of the planets; nay, and even to the remotest aphelion of the comets, if those aphelions are also quiescent. But hitherto I have not been able to discover the cause of those properties of gravity from phenomena, and I frame no hypotheses; for whatever is not deduced from the phenomena is to be called a hypothesis, and hypotheses, whether metaphysical or physical, whether of occult qualities or mechanical, have no place in experimental philosophy. In this philosophy particular propositions are inferred from the phenomena and afterward rendered general by induction. Thus it was that the impenetrability, the mobility, and the impulsive force of bodies, and the laws of motion and of gravitation, were discovered. And to us it is enough that gravity does really exist and act according to the laws which we have explained, and abundantly serves to account for all the motions of the celestial bodies and of our sea.

And now we might add something concerning a certain most subtle spirit which pervades and lies hid in all gross bodies, by the force and action of which spirit the particles of bodies attract one another at near distances and cohere, if contiguous; and electric bodies operate to greater distances, as well repelling as attracting the neighboring corpuscles; and light is emitted, reflected, refracted, inflected, and heats bodies; and all sensation is excited, and the members of animal bodies move at the command of the will, namely, by the vibrations of this spirit, mutually propagated along the solid filaments of the nerves, from the outward organs of sense to the brain and from the brain into the muscles. But these are things that cannot be explained in few words; nor are we furnished with that sufficiency of experiments which is required to an accurate determination and demonstration of the laws by which this electric and elastic spirit operates.

2. GOD AND GRAVITY [5]

FOUR LETTERS TO RICHARD BENTLEY [e]

I

To the Reverend Dr. Richard Bentley, at the Bishop of Worcester's House, in Park Street, Westminster

Sir,

When I wrote my treatise about our system, I had an eye upon such principles as might work with considering men for the belief of a Deity; and nothing can rejoice me more than to find it useful for that purpose. But if I have done the public any service this way, it is due to nothing but industry and patient thought.

As to your first query, it seems to me that if the matter of our sun and planets and all the matter of the universe were evenly scattered throughout all the heavens, and every particle had an innate gravity toward all the rest, and the whole space throughout which this matter was scattered was but finite, the matter on the outside of this space would, by its gravity, tend toward all the matter on the inside and, by consequence, fall down into the middle of the whole space and there compose one great spherical mass. But if the matter was evenly disposed throughout an infinite space, it could never convene into one mass; but some of it would convene into one mass and some into another, so as to make an infinite

---

[e] [*Opera Omnia* IV, pp. 429-42.]

number of great masses, scattered at great distances from one to another throughout all that infinite space. And thus might the sun and fixed stars be formed, supposing the matter were of a lucid nature. But how the matter should divide itself into two sorts, and that part of it which is fit to compose a shining body should fall down into one mass and make a sun and the rest which is fit to compose an opaque body should coalesce, not into one great body, like the shining matter, but into many little ones; or if the sun at first were an opaque body like the planets or the planets lucid bodies like the sun, how he alone should be changed into a shining body whilst all they continue opaque, or all they be changed into opaque ones whilst he remains unchanged, I do not think explicable by mere natural causes, but am forced to ascribe it to the counsel and contrivance of a voluntary Agent.

The same Power, whether natural or supernatural, which placed the sun in the center of the six primary planets, placed Saturn in the center of the orbs of his five secondary planets and Jupiter in the center of his four secondary planets, and the earth in the center of the moon's orb; and therefore, had this cause been a blind one, without contrivance or design, the sun would have been a body of the same kind with Saturn, Jupiter, and the earth, that is, without light and heat. Why there is one body in our system qualified to give light and heat to all the rest, I know no reason but because the Author of the system thought it convenient; and why there is but one body of this kind, I know no reason but because one was sufficient to warm and enlighten all the rest. For the Cartesian hypothesis of suns losing their light and then turning into comets, and comets into planets, can have no place in my system and is plainly erroneous; because it is certain that, as often as they appear to us, they descend into the system of our planets, lower than the orb of Jupiter and sometimes lower than the orbs of Venus and Mercury, and yet never stay here, but always return from the sun with the same degrees of motion by which they approached him.

To your second query, I answer that the motions which the planets now have could not spring from any natural cause alone, but were impressed by an intelligent Agent. For since comets descend into the region of our planets and here move all manner of ways, going sometimes the same way with the planets, sometimes the contrary way, and sometimes in crossways, in planes inclined to the plane of the ecliptic and at all kinds of angles, it is plain that there is no natural cause which could determine all the planets, both primary and secondary, to move the same way and in the same plane, without any considerable variation; this must have been the effect of counsel. Nor is there any natural cause which could give the planets those just degrees of velocity, in proportion to their distances from the sun and other central bodies, which were requisite to make them move in such concentric orbs about those bodies. Had the planets been as swift as comets, in proportion to their distances from the sun (as they would have been had their motion been caused by their gravity, whereby the matter, at the first formation of the planets, might fall from the remotest regions toward the sun), they would not move in concentric orbs, but in such eccentric ones as the comets move in. Were all the planets as swift as Mercury or as slow as Saturn or his satellites, or were their several velocities otherwise much greater or less than they are, as they might have been had they arose from any other cause than their gravities, or had the distances from the centers about which they move been greater or less than they are, with the same velocities, or had the quantity of matter in the sun or in Saturn, Jupiter, and the earth, and by consequence their gravitating power, been greater or less than it is, the primary planets could not have revolved about the sun nor the secondary ones about Saturn, Jupiter, and the earth, in concentric circles, as they do, but would have moved in hyperbolas or parabolas or in ellipses very eccentric. To make this system, therefore, with all its motions, required a cause which understood and compared together the quantities of matter in the several bodies of the sun and planets and the gravitating powers resulting from thence, the several distances of the primary planets from the sun and of the secondary ones from Saturn, Jupiter, and the earth, and the velocities with which these planets could revolve about those quantities of matter in the central bodies; and to compare and adjust all these things together, in so great a variety of bodies, argues that cause to be, not blind and fortuitous, but very well skilled in mechanics and geometry.

To your third query, I answer that it may be represented that the sun may, by heating those planets most of which are nearest to him, cause them to be better concocted and more condensed by that concoction. But when I consider that our earth is much more heated in its bowels below the upper crust by subterraneous fermentations of mineral bodies than by the sun, I see not why the interior parts of Jupiter and Saturn might not be as much heated, concocted, and coagulated by those fermentations as our earth is; and therefore this various density should have some other cause than the various distances of the planets from the sun. And I am confirmed in this opinion by considering that the planets of Jupiter and Saturn, as they are rarer than the rest, so they are vastly greater and contain a far greater quantity of matter, and have many satellites about them; which qualifications surely arose, not from their being placed at so great a distance from the sun, but were rather the cause why the Creator placed them at great distance. For, by their gravitating powers, they disturb one another's motions very sensibly, as I find by some late observations of Mr. Flamsteed; and had they been placed much nearer to the sun and to one another, they would, by the same powers, have caused a considerable disturbance in the whole system.

To your fourth query, I answer that, in the hypothesis of vortices, the inclination of the axis of the earth might, in my opinion, be ascribed to the situation of the earth's vortex before it was absorbed by the neighboring vortices and the earth turned from a sun to a comet; but this inclination ought to decrease constantly in compliance with the motion of the earth's vortex, whose axis is much less inclined to the ecliptic, as appears by the motion of the moon carried about therein. If the sun by his rays could carry about the planets, yet I do not see how he could thereby effect their diurnal motions.

Lastly, I see nothing extraordinary in the inclination of the earth's axis for proving a Deity, unless you will urge it as a contrivance for winter and summer, and for making the earth habitable toward the poles; and that the diurnal rotations of the sun and planets, as they could hardly arise from any cause purely mechanical, so by being determined all the same way with the annual and menstrual motions they seem to make up that harmony in the system which, as I explained above, was the effect of choice rather than chance.

There is yet another argument for a Deity, which I take to be a very strong one; but till the principles on which it is grounded are better received, I think it more advisable to let it sleep.

I am your most humble servant to command,

Is. NEWTON

Cambridge, December 10, 1692

II

For Mr. Bentley, at the Palace at Worcester

Sir,

I agree with you that if matter evenly diffused through a finite space, not spherical, should fall into a solid mass, this mass would affect the figure of the whole space, provided it were not soft, like the old chaos, but so hard and solid from the beginning that the weight of its protuberant parts could not make it yield to their pressure; yet, by earthquakes loosening the parts of this solid, the protuberances might sometimes sink a little by their weight, and thereby the mass might by degrees approach a spherical figure.

The reason why matter evenly scattered through a finite space would convene in the midst you conceive the same with me, but that there should be a central particle so accurately placed in the middle as to be always equally attracted on all sides, and thereby continue without motion, seems to me a supposition fully as hard as to make the sharpest needle stand upright on its point upon a looking glass. For if the very mathematical center of the central particle be not accurately in the very mathematical center of the attractive power of the whole mass, the particle will not be attracted equally on all sides. And much harder it is to suppose all the particles in an infinite space should be so accurately poised one among another as to stand still in a perfect equilibrium. For I reckon this as hard as to make, not one needle only, but an infinite number of them (so many as there are particles in an infinite space) stand accurately poised upon their points. Yet I grant it possible, at least by a divine power; and if they were once to be placed, I agree with you that they would continue in that posture without motion forever, unless put into new motion by the same power. When, therefore, I said that matter evenly spread through all space would convene by its gravity into one or more great masses, I understand it of matter not resting in an accurate poise.

But you argue, in the next paragraph of your letter, that every particle of matter in an infinite space has an infinite quantity of matter on all sides and, by consequence, an infinite attraction every way, and therefore must rest *in equilibrio,* because all infinites are equal. Yet you suspect a paralogism in this argument, and I conceive the paralogism lies in the position that all infinites are equal. The generality of mankind consider infinites no other ways than indefinitely; and in this sense they say all infinites are equal, though they would speak more truly if they should say they are neither equal nor unequal, nor have any certain difference or proportion one to another. In this sense, therefore, no conclusions can be drawn from them about the equality, proportions, or differences of things; and they that attempt to do it usually fall into paralogisms. So when men argue against the infinite divisibility of magnitude by saying that if an inch may be divided into an infinite number of parts the sum of those parts will be an inch; and if a foot may be divided into an infinite number of parts the sum of those parts must be a foot; and therefore, since all infinites are equal, those sums must be equal, that is, an inch equal to a foot.

The falseness of the conclusion shows an error in the premises, and the error lies in the position that all infinites are equal. There is, therefore, another way of considering infinites used by mathematicians, and that is, under certain definite restrictions and limitations, whereby infinites are determined to have certain differences or proportions to one another. Thus Dr. Wallis considers them in his *Arithmetica Infinitorum,* where, by the various proportions of infinite sums, he gathers the various proportions of infinite magnitudes, which way of arguing is generally allowed by mathematicians and yet would not be good were all infinites equal. According to the same way of considering infinites, a mathematician would tell you that, though there be an infinite number of infinite little parts in an inch, yet there is twelve times that number of such parts in a foot; that is, the infinite number of those parts in a foot is not equal to but twelve times bigger than the infinite number of them in an inch. And so a mathematician will tell you that if a body stood *in equilibrio* between any two equal and contrary attracting infinite forces, and if to either of these forces you add any new finite attracting force, that new force, howsoever little, will destroy their equilibrium and put the body into the same motion into which it would put it were those two contrary equal forces but finite or even none at all; so that in this case the two equal infinites, by the addition of a finite to either of them, become unequal in our ways of reckoning; and after these ways we must reckon, if from the considerations of infinites we would always draw true conclusions.

To the last part of your letter, I answer, first, that if the earth (without the moon) were placed anywhere with its center in the *orbis magnus* and stood still there without any gravitation or projection, and there at once were infused into it both a gravitating energy toward the sun and a transverse impulse of a just quantity moving it directly in a tangent to the *orbis magnus*, the compounds of this attraction and projection would, according to my notion, cause a circular revolution of the earth about the sun. But the transverse impulse must be a just quantity; for if it be too big or too little, it will cause the earth to move in some other line. Secondly, I do not know any power in nature which would cause this transverse motion without the divine arm. Blondel tells us somewhere in his book of Bombs that Plato affirms that the motion of the planets is such as if they had all of them been created by God in some region very remote from our system and let fall from thence toward the sun, and so soon as they arrived at their several orbs their motion of falling turned aside into a transverse one. And this is true, supposing the gravitating power of the sun was double at that moment of time in which they all arrive at their several orbs; but then the divine power is here required in a double respect, namely, to turn the descending motions of the falling planets into a side motion and, at the same time, to double the attractive power of the sun. So, then, gravity may put the planets into motion, but without the divine power it could never put them into such a circulating motion as they have about the sun; and therefore, for this as well as other reasons, I am compelled to ascribe the frame of this system to an intelligent Agent.

You sometimes speak of gravity as essential and inherent to matter. Pray do not ascribe that notion to me, for the cause of gravity is what I do not pretend to know and therefore would take more time to consider of it.

I fear what I have said of infinites will seem obscure to you; but it is enough if you understand that infinites, when considered absolutely without any restriction or limitation, are neither equal nor unequal, nor have any certain proportion one to another, and therefore the principle that all infinites are equal is a precarious one.

Sir, I am your most humble servant,

Is. NEWTON

Trinity College, January 17, 1692/3

III

For Mr. Bentley, at the Palace at Worcester

Sir,

Because you desire speed, I will answer your letter with what brevity I can. In the six positions you lay down in the beginning of your letter, I agree with you. Your assuming the *orbis magnus* 7,000 diameters of the earth wide implies the sun's horizontal parallax to be half a minute. Flamsteed and Cassini have of late observed it to be about 10 minutes, and thus the *orbis magnus* must be 21,000, or, in a round number, 20,000 diameters of the earth wide. Either computation, I think, will do well; and I think it not worth while to alter your numbers.

In the next part of your letter you lay down four other positions, founded upon the six first. The first of these four seems very evident, supposing you take attraction so generally as by it to understand any force by which distant bodies endeavor to come together without mechanical impulse. The second seems not so clear, for it may be said that there might be other systems of worlds before the present ones and others before those, and so on to all past eternity, and by consequence that gravity may be coeternal to matter and have the same effect from all eternity as at present, unless you have somewhere proved that old systems cannot gradually pass into new ones or that this system had not its original from the exhaling matter of former decaying systems but from a chaos of matter evenly dispersed throughout all space; for something of this kind, I think you say, was the subject of your Sixth Sermon, and the growth of new systems out of old ones, without the mediation of a divine power, seems to me apparently absurd.

The last clause of the second position I like very well. It is inconceivable that inanimate brute matter should, without the mediation of something else which is not material, operate upon and affect other matter without mutual contact, as it must be if gravitation, in the sense of Epicurus, be essential and inherent in it. And this is one reason why I desired you would not ascribe innate gravity to me. That gravity should be innate, inherent, and essential to matter, so that one body may act upon another at a distance through a *vacuum,* without the mediation of anything else, by and through which their action and force may be conveyed from one to another, is to me so great an absurdity that I believe no man who has in philosophical matters a competent faculty of thinking can ever fall into it. Gravity must be caused by an agent acting constantly according to certain laws, but whether this agent be material or immaterial I have left to the consideration of my readers.

Your fourth assertion, that the world could not be formed by innate gravity alone, you confirm by three arguments. But in your first argument you seem to make a *petitio principii;* for whereas many ancient philosophers and others, as well theists as atheists, have all allowed that there may be worlds and parcels of matter innumerable or infinite, you deny this by representing it as absurd as that there should be positively an infinite arithmetical sum or number, which is a contradiction *in terminis*, but you do not prove it as absurd. Neither do you prove that what men mean by an infinite sum or number is a contradiction in nature, for a contradiction *in terminis* implies no more than an impropriety of speech. Those things which men understand by improper and contradictious phrases may be sometimes really in nature without any contradiction at all: a silver inkhorn, a paper lantern, an iron whetstone, are absurd phrases, yet the things signified thereby are really in nature. If any man should say that a number and a sum, to speak properly, is that which may be numbered and summed, but things infinite are numberless or, as we usually speak, innumerable and sumless or insummable, and therefore ought not to be called a number or sum, he will speak properly enough, and your argument against him will, I fear, lose its force. And yet if any man shall take the words 'number' and 'sum' in a larger sense, so as to understand thereby things which, in the proper way of speaking, are numberless and sumless (as you seem to do when you allow an infinite number of points in a line), I could readily allow him the use of the contradictious phrases of 'innumerable number' or 'sumless sum,' without inferring from thence any absurdity in the thing he means by those phrases. However, if by this or any other argument you have proved the finiteness of the universe, it follows that all matter would fall down from the outsides and convene in the middle. Yet the matter in falling might concrete into many round masses, like the bodies of the planets, and these, by attracting one another, might acquire an obliquity of descent by means of which they might fall, not upon the great central body, but upon the side of it, and fetch a compass about and then ascend again by the same steps and degrees of motion and velocity with which they descended before, much after the manner that comets revolve about the sun; but a circular motion in concentric orbs about the sun they could never acquire by gravity alone.

And though all the matter were divided at first into several systems, and every system by a divine power constituted like ours, yet would the outside systems descend toward the middlemost; so that this frame of things could not always subsist without a divine power to conserve it, which is the second argument; and to your third I fully assent.

As for the passage of Plato, there is no common place from whence all the planets, being let fall and descending with uniform and equal gravities (as Galileo supposes), would, at their arrival to their several orbs, acquire their several velocities with which they now revolve in them. If we suppose the gravity of all the planets toward the sun to be of such a quantity as it really is, and that the motions of the planets are turned upward, every planet will ascend to twice its height from the sun. Saturn will ascend till he be twice as high from the sun as he is at present, and no higher; Jupiter will ascend as high again as at present, that is, a little above the orb of Saturn; Mercury will ascend to twice his present height, that is, to the orb of Venus; and so of the rest; and then, by falling down again from the places to which they ascended, they will arrive again at their several orbs with the same velocities they had at first and with which they now revolve.

But if, so soon as their motions by which they revolve are turned upward, the gravitating power of the sun, by which their ascent is perpetually retarded, be diminished by one half, they will now ascend perpetually, and all of them at equal distances from the sun will be equally swift. Mercury, when he arrives at the orb of Venus, will be as swift as Venus; and he and Venus, when they arrive at the orb of the earth, will be as swift as the earth; and so of the rest. If they begin all of them to ascend at once and ascend in the same line, they will constantly, in ascending, become nearer and nearer together, and their motions will constantly approach to an equality and become at length slower than any motion assignable. Suppose, therefore, that they ascended till they were almost contiguous and their motions inconsiderably little, and that all their motions were at the same moment of time turned back again or, which comes almost to the same thing, that they were only deprived of their motions and let fall at that time; they would all at once arrive at their several orbs, each with the velocity it had at first, and if their motions were then turned sideways and at the same time the gravitating power of the sun doubled, that it might be strong enough to retain them in their orbs, they would revolve in them as before their ascent. But if the gravitating power of the sun was not doubled, they would go away from their orbs into the highest heavens in parabolical lines. These things follow from my *Principia Mathematica*, Book I, Propositions XXXIII, XXXIV, XXXVI, XXXVII.

I thank you very kindly for your designed present, and rest

Your most humble servant to command,

Is. NEWTON

Cambridge, February 25, 1692/3

IV

For Mr. Bentley, at the Palace at Worcester

Sir,

The hypothesis of deriving the frame of the world by mechanical principles from matter evenly spread through the heavens being inconsistent with my system, I had considered it very little before your letters put me upon it, and therefore trouble you with a line or two more about it, if this comes not too late for your use.

In my former I represented that the diurnal rotations of the planets could not be derived from gravity, but required a divine arm to impress them. And though gravity might give the planets a motion of descent toward the sun, either directly or with some little obliquity, yet the transverse motions by which they revolve in their several orbs required the divine arm to impress them according to the tangents of their orbs. I would now add that the hypothesis of matters being at first evenly spread through the heavens is, in my opinion, inconsistent with the hypothesis of innate gravity, without a supernatural power to reconcile them; and therefore it infers a Deity. For if there be innate gravity, it is impossible now for the matter of the earth and all the planets and stars to fly up from them, and become evenly spread throughout all the heavens, without a supernatural power; and certainly that which can never be hereafter without a supernatural power could never be heretofore without the same power.

You queried whether matter evenly spread throughout a finite space, of some other figure than spherical, would not, in falling down toward a central body, cause that body to be of the same figure with the whole space, and I answered yes. But in my answer it is to be supposed that the matter descends directly downward to that body and that that body has no diurnal rotation. This, sir, is all I would add to my former letters.

I am your most humble servant,

Is. NEWTON

Cambridge, February 11, 1693

3. ON CREATION [6]

From a letter to Thomas Burnet [f]

... You seem to apprehend that I would have the present face of the earth formed in the first creation. A sea I believe was then formed, as Moses expresses, but not like the sea, but with an even bottom without any precipices or steep descents, as I think I expressed in my letter. Of our present sea, rocks, mountains, etc., I think you have given the most plausible account. And yet if one would go about to explain it otherwise, philosophically, he might say that as saltpeter dissolved in water, though the solution be uniform, crystallizes not all over the vessel alike, but here and there in long bars of salt; so the limus of the chaos, or some substances in it, might coagulate at first, not all over the earth alike, but here and there in veins or beds of divers sorts of stones and minerals. That in other places which remained yet soft the air, which in some measure subsided out of the superior regions of the chaos, together with the earth or limus by degrees extricating itself gave liberty to the limus to shrink and subside, and leave the first coag-

---

[f] [From an undated copy in Newton's hand, probably written in 1681. Quoted in Brewster, *Memoirs of Sir Isaac Newton*, Vol. II, pp. 447-54; see also pp. 99-100.]

ulated places standing up like hills, which subsiding would be increased by the draining and drying of that limus. That the veins and tracts of limus in the bowels of those mountains also drying, and consequently shrinking, cracked and left many cavities, some dry, others filled with water. That after the upper crust of the earth, by the heat of the sun together with that caused by action of minerals, had hardened and set, the earth in the lower regions still going closer together left large caverns between it, and the upper crust filled with the water, which, upon subsiding by its weight, it spread out by degrees till it had done shrinking, which caverns or subterraneal seas might be the great deep of Moses and, if you will, it may be supposed one great orb of water between the upper crust or gyrus and the lower earth, though perhaps not a very regular one. That in process of time many exhalations were gathered in those caverns which would have expanded themselves into 40 or 50 times the room they lay in, or more, had they been at liberty. For if air in a glass may be crowded into 18 or 20 times less room than it takes at liberty, and yet not burst the glass, much more may subterranean exhalations by the vast weight of the incumbent earth be kept crowded into a less room before they can in any place lift up and burst that crust of earth. That at length somewhere forcing a breach, they, by expanding themselves, forced out vast quantities of water before they could all get out themselves, which commotion caused tempests in the air and thereby great falls of rain in spouts, and all together made the flood; and after the vapors were out, the waters retired into their former place. That the air which in the beginning subsided with the earth, by degrees extricating itself, might be pent up in one or more great caverns in the lower earth under the abyss, and at the time of the flood, breaking out into the abyss and consequently expending itself, might also force out the waters of the abyss before it. That the upper crust or gyrus of earth might be upon the stretch before the breaking out of the abyss and then, by its weight shrinking to its natural posture, might help much to force out the waters. That the subterraneal vapors which then first broke out and have ever since continued to do so, being found by experience noxious to man's health, infect the air and cause that shortness of life which has been ever since the flood. And that several pieces of earth, either at the flood or since falling, some perhaps into the great deep, others into less and shallower cavities, have caused many of those phenomena we see on the earth, besides the original hills and cavities.

But, you will ask, how could a uniform chaos coagulate at first irregularly in heterogeneous veins or masses to cause hills? Tell me then how a uniform solution of saltpeter coagulates irregularly into long bars; or, to give you another instance, if tin (such as the pewterers bring from the mines in Cornwel to make pewter of) be melted and then let stand to cool till it begin to congeal, and when it begins to congeal at the edges, if it be inclined on one side for the more fluid part of the tin to run from those parts which coagulate first, you will see a good part of the tin congealed in lumps, which, after the more fluid part of the tin which congeals not so soon is run from between them, appear like so many hills, with as much irregularity as any hills on the earth do. Tell me the cause of this, and the answer will perhaps serve for the chaos.

All this I write not to oppose you, for I think the main part of your hypothesis as probable as that I have here written, if not in some respects more probable. And though the pressure of the moon or vortex, etc., may promote the irregularity of the causes of hills, yet I did not in my former letter design to explain the generation of hills thereby, but only to insinuate how a sea might be made above ground in your own hypothesis before the flood, besides the subterranean great deep, and thereby all difficulty of explaining rivers and the main point in which some may think you and Moses disagree might be avoided. But this sea I [do] not suppose round the equator, but rather to be two seas in two opposite parts of it where the cause of the flux and reflux of our present sea depressed the soft mass of the earth at that time when the upper crust of it hardened.

As to Moses, I do not think his description of the creation either philosophical or feigned, but that he described realities in a language artificially adapted to the sense of the vulgar. Thus when he speaks of two great lights, I suppose he means their apparent, not real, greatness. So when he tells us God placed these lights in the firmament, he speaks I suppose of their apparent, not real, place, his business being, not to correct the vulgar notions in matters philosophical, but to adapt a description of the creation as handsomely as he could to the sense and capacity of the vulgar. So when he tells us of two great lights and the stars made the fourth day, I do not think their creation from beginning to end was done the fourth day nor in any one day of the creation, nor that Moses mentions their creation as they were physical bodies in themselves, some of them greater than this earth and perhaps habitable worlds, but only as they were lights to this earth; so therefore, though their creation could not physically [be] assigned to any one day, yet being a part of the sensible creation which it was Moses's design to describe and it being his design to describe things in order according to the succession of days, allotting no more than one day to one thing, they were to be referred to some day or other, and rather to the fourth day than any other, if [the] air then first became clear enough for them to shine through it, and so put on the appearances of lights in the firmament to enlighten the earth. For till then they could not properly be described under the notion of such lights; nor was their description under that notion to be deferred after they had that appearance, though it may be the creation of some of them was not yet completed. Thus far, perhaps, one might be allowed to go in the explaining [of] the creation of the fourth day, but in the third day for Moses to describe the creation of seas when there was no such thing done, neither in reality nor appearance, methinks is something hard; and that the rather because if before the flood there was no water but that of rivers, that is, none but fresh water above ground, there could be no fish but such as live in fresh water, and so one half of the fifth day's work will be a nonentity, and God must be put upon a creation after the flood to replenish one half of this terraqueous globe with whales and all those other kinds of sea fish we now have.

You ask what was that light created the first day? Of what extent was the Mosaical chaos? Was the firmament, if taken for the atmosphere, so considerable a thing as to take up one day's work, and would not the description of the creation have been complete without mentioning it? To answer these things fully would require comment upon Moses, whom I dare not pretend to understand; yet to say something by way of conjecture, one may suppose that all the planets about our sun were created together, there being in no history any mention of new ones appearing or old ones ceasing. That they all, and the sun too, had at first one common chaos. That this chaos, by the spirit of God moving upon it, became separated into several parcels, each parcel for a planet. That at the same time the matter of the sun also separated from the rest, and upon the separation began to shine before it was formed into that compact and well-defined body we now see it. And the preceding darkness and light now cast upon the chaos of every planet from the solar chaos was the evening and morning, which Moses calls the first day, even before the earth had any diurnal motion or was formed into a globular body. That it being Moses' design to describe the origination of this earth only and to touch upon other things only so far as they [be] related to it, he passes over the division of the general chaos into particular ones and does not so much as describe the fountain of that light God made, that is, the chaos of the sun, but only with respect to the chaos of the earth tells us that God made light upon the face of the deep where darkness was before. Further, one might suppose that after the chaos was separated from the rest, by the same principle which promoted its separation (which might be gravitation toward a center), it shrunk closer together and at length, a great part of it condensing, subsided in the form of a muddy water or limus to compose this terraqueous globe. The rest which condensed not separated into two parts, the vapors above and the air, which being of a middle degree of gravity ascended from the one, descended from the other, and gathered into a body stagnating between both. Thus was the chaos at once separated into three regions, the globe of muddy waters below the firmament, the vapors or waters above the firmament, and the air or firmament itself. Moses had before called the chaos "the deep" and "the waters," on the face of which the spirit of God moved, and here he teaches the division of all those waters into two parts, with a firmament between them, which being the main step in the generation of this earth was in no wise to be omitted by Moses. After this general division of the chaos, Moses teaches a subdivision of one of its parts, that is, of the miry waters under the firmament into clear water and dry land on the surface of the whole globous mass, for which separation nothing more was requisite than that the water should be drained from the higher parts of the limus to leave them dry land and gather together into the lower to compose seas. And some parts might be made higher than others, not only by the cause of the flux and reflux, but also by the figure of the chaos, if it was made by division from the chaos of other planets; for then it could not be spherical. And now while the new planted vegetables grew to be food for animals, the heavens becoming clear for the sun in the day and moon and stars in the night to shine distinctly through them on the earth, and so put on the form of lights in the firmament, so that had men been now living on the earth to view the process of the creation they would have judged those lights created at this time. Moses here sets down their creation as if he had then lived and were now describing what he saw. Omit them he could not without rendering his description of the creation imperfect in the judgment of the vulgar. To describe them distinctly as they were in themselves would have made the narration tedious and confused, amused the vulgar, and become a philosopher more than a prophet. He mentions them, therefore, only so far as the vulgar had a notion of them, that is, as they were phenomena in the firmament, and describes their making only so far and at such a time as they were made such phenomena. Consider, therefore, whether anyone who understood the process of the creation and designed to accommodate to the vulgar not an ideal or poetical but a true description of it, as succinctly and theologically as Moses has done, without omitting anything material which the vulgar have a notion of or describing any being further than the vulgar have a notion of it, could mend that description which Moses has given us. If it be said that the expression of making and setting two great lights in the firmament is more poetical than natural, so also are some other expressions of Moses, as when he tells us the windows or floodgates of heaven were opened (Genesis 7) and afterward stopped again (Genesis 8), and yet the things signified by such figurative expressions are not ideal or moral but true. For Moses, accommodating his words to the gross conceptions of the vulgar, describes things much after the manner as one of the vulgar would have been inclined to do had he lived and seen the whole series of what Moses describes.

Now for the number and length of the six days: by what is said above, you may make the first day as long as you please and the second day too, if there was no diurnal motion till there was a terraqueous globe, that is, till toward the end of that day's work. And then if you will suppose the earth put in motion by an even force applied to it and that the first revolution was done in one of our years, in the time of another year there would be three revolutions, of a third five, of a fourth seven, etc., and of the 183rd year 365 revolutions, that is, as many as there are days in our year; and, in all this time, Adam's life would be increased but about 90 of our years, which is no such great business. But yet I must profess I know no sufficient natural cause of the earth's diurnal motion. Where natural causes are at hand, God uses them as instruments in his works; but I do not think them alone sufficient for the creation, and therefore may be allowed to suppose that, amongst other things, God gave the earth its motion by such degrees and at such times as was most suitable to the creatures. If you would have a year for each day's work, you may, by supposing day and night was made by the annual motion of the earth only and that the earth had no diurnal motion till toward the end of the six days. But you will complain of long and doleful nights; and why might not birds and fishes endure one long night as well as those and many other animals endure many in Greenland; or rather why not better than the tender substances which were growing into animals might endure successions of short days and nights, and consequently of heat and cold? For what think you would become of an egg or embryo which should frequently grow hot and cold? Yet if you think the night too long, it is but supposing the divine operations quicker. But be it as it will, methinks one of the Ten Commandments given by God in Mount Sinai, pressed by divers of the prophets, observed by our Savior, his

Apostles, and first Christians for 300 years, and with a day's alteration by all Christians to this day, should not be grounded on a fiction. At least divines will hardly be persuaded to [be]lieve so.

As I am writing, another illustration of the generation of hills, proposed above, comes into my mind. Milk is as uniform a liquor as the chaos was. If beer be poured into it and the mixture let stand till it be dry, the surface of the curdled substance will appear as rugged and mountainous as the earth in any place. I forbear to describe other causes of mountains, as the breaking out of vapors from below before the earth was well hardened, the settling and shrinking of the whole globe after the upper regions or surface began to be hard. Nor will I urge their antiquity out of Proverbs 8:25, Job 15:7, Psalm 90:2, but rather beg your excuse for this tedious letter, which I have the more reason to do because I have not set down anything I have well considered or will undertake to defend.

4. ON UNIVERSAL DESIGN [7]

From a Manuscript [g]

Opposite to godliness is atheism in profession and idolatry in practice. Atheism is so senseless and odious to mankind that it never had many professors. Can it be by accident that all birds, beasts, and men have their right side and left side alike shaped (except in their bowels); and just two eyes, and no more, on either side of the face; and just two ears on either side [of] the head; and a nose with two holes; and either two forelegs or two wings or two arms on the shoulders, and two legs on the hips, and no more? Whence arises this uniformity in all their outward shapes but from the counsel and contrivance of an Author? Whence is it that the eyes of all sorts of living creatures are transparent to the very bottom, and the only transparent members in the body, having on the outside a hard transparent skin

g ["A Short Scheme of the True Religion." Quoted in Brewster, *op. cit.*, Vol. II, pp. 347-8.]

and within transparent humors, with a crystalline lens in the middle and a pupil before the lens, all of them so finely shaped and fitted for vision that no artist can mend them? Did blind chance know that there was light and what was its refraction, and fit the eyes of all creatures after the most curious manner to make use of it? These and suchlike considerations always have and ever will prevail with mankind to believe that there is a Being who made all things and has all things in his power, and who is therefore to be feared. . . .

We are, therefore, to acknowledge one God, infinite, eternal, omnipresent, omniscient, omnipotent, the Creator of all things, most wise, most just, most good, most holy. We must love him, fear him, honor him, trust in him, pray to him, give him thanks, praise him, hallow his name, obey his commandments, and set times apart for his service, as we are directed in the Third and Fourth Commandments, for this is the love of God that we keep his commandments, and his commandments are not grievous (I John 5:3). And these things we must do not to any mediators between him and us, but to him alone, that he may give his angels charge over us, who, being our fellow servants, are pleased with the worship which we give to their God. And this is the first and the principal part of religion. This always was and always will be the religion of all God's people, from the beginning to the end of the world.

From a Manuscript [h]

God made and governs the world invisibly and has commanded us to love and worship him, and no other God; to honor our parents and masters, and love our neighbors as ourselves; and to be temperate, just, and peaceable; and to be merciful even to brute beasts. And by the same power by which he gave life at first to every species of animals he is able to revive the dead, and has revived Jesus Christ our Redeemer, who has gone into the heavens

---

[h] [Quoted in Brewster, *op. cit.*, Vol. II, p. 354. I have not been able to ascertain whether the italics at the end are indicated in the manuscript or are due to Brewster's enthusiasm for this passage.]

to receive a kingdom and prepare a place for us, and is next in dignity to God and may be worshiped as the Lamb of God, and has sent the Holy Ghost to comfort us in his absence, and will at length return and reign over us, invisibly to mortals, till he has raised up and judged all the dead; and then he will give up his kingdom to the Father and carry the blessed to the place he is now preparing for them, and send the rest to other places suitable to their merits. *For in God's house (which is the universe) are many mansions, and he governs them by agents which can pass through the heavens from one mansion to another. For if all places to which we have access are filled with living creatures, why should all these immense spaces of the heavens above the clouds be incapable of inhabitants?*

IV. Questions on Natural Philosophy

1. THE NEW THEORY ABOUT LIGHT AND COLORS [8]

Communicated to the Royal Society, February 6, 1671/2 [a]

Sir,

To perform my late promise to you, I shall without further ceremony acquaint you that in the beginning of the year 1666 (at which time I applied myself to the grinding of optic glasses of other figures than spherical) I procured me a triangular glass prism, to try therewith the celebrated phenomena of colors. And in order thereto having darkened my chamber and made a small hole in my window shuts, to let in a convenient quantity of the sun's light, I placed my prism at his entrance, that it might be thereby refracted to the opposite wall. It was at first a pleasing divertisement to view the vivid and intense colors produced thereby; but after a while applying myself to consider them more circumspectly, I became surprised to see them in an *oblong* form; which, according to the received laws of refraction, I expected should have been *circular*.

They were terminated at the sides with straight lines, but at the ends the decay of light was so gradual that it was difficult to determine justly what was their figure; yet they seemed semicircular.

Comparing the length of this colored spectrum with its breadth, I found it about five times greater, a disproportion so extravagant that it excited me to a more than ordinary curiosity of examining from whence it might proceed. I could scarce think that the various thickness of the glass, or the termination with shadow or darkness, could have any influence on light to produce

---

[a] [Philosophical Transactions of the Royal Society, No. 80, Feb. 19, 1672, pp. 3075-87. Also given, with some revisions and minor deletions, in *Opera Omnia* IV, pp. 295-308.]

68

such an effect; yet I thought it not amiss first to examine those circumstances, and so tried what would happen by transmitting light through parts of the glass of divers thicknesses or through holes in the window of divers bignesses, or by setting the prism without so that the light might pass through it and be refracted before it was terminated by the hole; but I found none of those circumstances material. The fashion of the colors was in all these cases the same.

Then I suspected whether, by any unevenness in the glass or other contingent irregularity, these colors might be thus dilated. And to try this I took another prism like the former and so placed it that the light, passing through them both, might be refracted contrary ways, and so by the latter returned into that course from which the former had diverted it. For by this means I thought the regular effects of the first prism would be destroyed by the second prism, but the irregular ones more augmented by the multiplicity of refractions. The event was that the light, which by the first prism was diffused into an oblong form, was by the second reduced into an orbicular one, with as much regularity as when it did not at all pass through them. So that, whatever was the cause of that length, it was not any contingent irregularity.

I then proceeded to examine more critically what might be effected by the difference of the incidence of rays coming from divers parts of the sun; and to that end measured the several lines and angles belonging to the image. Its distance from the hole or prism was 22 feet; its utmost length $13\frac{1}{4}$ inches; its breadth $2\frac{5}{8}$; the diameter of the hole $\frac{1}{4}$ of an inch; the angle, with the rays tending toward the middle of the image, made with those lines, in which they would have proceeded without refraction, was 44 degrees, 56 minutes. And the vertical angle of the prism, 63 degrees, 12 minutes. Also the refractions on both sides [of] the prism, that is, of incident and emergent rays, were, as near as I could make them, equal, and consequently about 54 degrees, 4 minutes. And the rays fell perpendicularly upon the wall. Now subducting the diameter of the hole from the length and breadth of the image, there remains 13 inches the length and

$2\frac{1}{8}$ the breadth, comprehended by those rays which passed through the center of the said hole; and consequently the angle of the whole which that breadth subtended was about 31 minutes, answerable to the sun's diameter, but the angle which its length subtended was more than five such diameters, namely, 2 degrees, 49 minutes.

Having made these observations, I first computed from them the refractive power of that glass and found it measured 'by the ratio of the sines, 20 to 31. And then, by that ratio, I computed the refractions of two rays flowing from opposite parts of the sun's discus, so as to differ 31 minutes in their obliquity of incidence, and found that the emergent rays should have comprehended an angle of about 31 minutes, as they did, before they were incident.

But because this computation was founded on the hypothesis of the proportionality of the sines of incidence and refraction, which, though by my own experience I could not imagine to be so erroneous as to make that angle but 31 minutes which in reality was 2 degrees, 49 minutes, yet my curiosity caused me again to take my prism. And having placed it at my window as before, I observed that by turning it a little about its axis to and fro, so as to vary its obliquity to the light more than an angle of 4 or 5 degrees, the colors were not thereby sensibly translated from their place on the wall; and consequently, by that variation of incidence, the quantity of refraction was not sensibly varied. By this experiment therefore, as well as by the former computation, it was evident that the difference of the incidence of rays flowing from divers parts of the sun could not make them after decussation diverge at a sensibly greater angle than that at which they before converged, which being at most but about 31 or 32 minutes, there still remained some other cause to be found out from whence it could be 2 degrees 49 minutes.

Then I began to suspect whether the rays, after their trajection through the prism, did not move in curved lines, and according to their more or less curvity tend to divers parts of the wall. And it increased my suspicion when I remembered that I had often seen a tennis ball, struck with an oblique racket, describe such a curved line. For a circular as well as a progressive motion being communicated to it by that stroke, its parts on that side where the motions conspire must press and beat the contiguous air more violently than on the other, and there excite a reluctancy and reaction of the air proportionably greater. And for the same reason, if the rays of light should possibly be globular bodies,[b] and by their oblique passage out of one medium into another acquire a circulation motion, they ought to feel the greater resistance from the ambient ether on that side where the motions conspire, and thence be continually bowed to the other. But notwithstanding this plausible ground of suspicion, when I came to examine it I could observe no such curvity in them. And besides (which was enough for my purpose) I observed that the difference between the length of the image and diameter of the hole through which the light was transmitted was proportional to their distance.

The gradual removal of these suspicions at length led me to the *Experimentum Crucis,* which was this: I took two boards and placed one of them close behind the prism at the window, so that the light might pass through a small hole, made in it for the purpose, and fall on the other board, which I placed at about 12 feet distance, having first made a small hole in it also for some of that incident light to pass through. Then I placed another prism behind this second board, so that the light, trajected through both the boards, might pass through that also and be again refracted before it arrived at the wall. This done, I took the first prism in my hand and turned it to and fro slowly about its axis, so much as to make the several parts of the image cast on the second board successively pass through the hole in it, that I might observe to what places on the wall the second prism would refract them. And I saw by the variation of those places that the light, tending to that end of the image toward which the refraction of the first prism was made, did in the second prism suffer a refraction considerably greater than the light tending to the other end. And so the true cause of the length of that image was detected to be no other than that *light* consists of *rays differently refrangible,* which, without any respect to a difference in their incidence were, accord-

b [This was the Cartesian hypothesis.]

ing to their degrees of refrangibility, transmitted toward divers parts of the wall.

When I understood this I left off my aforesaid glass works, for I saw that the perfection of telescopes was hitherto limited, not so much for want of glasses truly figured according to the prescriptions of optic authors (which all men have hitherto imagined), as because that light itself is a *heterogeneous mixture of differently refrangible rays.* So that, were a glass so exactly figured as to collect any one sort of rays into one point, it could not collect those also into the same point which, having the same incidence upon the same medium, are apt to suffer a different refraction. Nay, I wondered that, seeing the difference of refrangibility was so great as I found it, telescopes should arrive to that perfection they are now at. For, measuring the refractions in one of my prisms, I found that, supposing the common sine of incidence upon one of its planes was 44 parts, the sine of refraction of the utmost rays on the red end of the colors made out of the glass into the air would be 68 parts, and the sine of refraction of the utmost rays on the other end 69 parts, so that the difference is about a 24th or 25th part of the whole refraction. And consequently the object glass of any telescope cannot collect all the rays which come from one point of an object so as to make them convene at its focus in less room than in a circular space whose diameter is the 50th part of the diameter of its aperture; which is an irregularity some hundreds of times greater than a circularly figured lens, of so small a section as the object glasses of long telescopes are, would cause by the unfitness of its figure, were light uniform.

This made me take reflections into consideration, and finding them regular, so that the angle of reflection of all sorts of rays was equal to their angle of incidence, I understood that by their mediation optic instruments might be brought to any degree of perfection imaginable, provided a reflecting substance could be found which would polish as finely as glass and reflect as much light as glass transmits, and the art of communicating to it a parabolic figure be also attained. But there seemed very great difficulties, and I have almost thought them insuperable when I further considered that every irregularity in a reflecting superficies makes the rays stray five or six times more out of their due course than the like irregularities in a refracting one; so that a much greater curiosity would be here requisite than in figuring glasses for refraction.

Amidst these thoughts I was forced from Cambridge by the intervening plague, and it was more than two years before I proceeded further. But then having thought of a tender way of polishing, proper for metal, whereby, as I imagined, the figure also would be corrected to the last, I began to try what might be effected in this kind, and by degrees so far perfected an instrument (in the essential parts of it like that I sent to London) by which I could discern Jupiter's four concomitants, and showed them divers times to two others of my acquaintance. I could also discern the moonlike phase of Venus, but not very distinctly, nor without some niceness in disposing the instrument.

From that time I was interrupted till this last autumn, when I made the other. And as that was sensibly better than the first (especially for day objects), so I doubt not but they will be still brought to a much greater perfection by their endeavors who, as you inform me, are taking care about it at London.

I have sometimes thought to make a microscope which in like manner should have, instead of an object glass, a reflecting piece of metal. And this I hope they will also take into consideration. For those instruments seem as capable of improvement as telescopes, and perhaps more, because but one reflective piece of metal is requisite in them, as you may perceive by the annexed diagram, where A B represents the object metal, C D the eyeglass, F their common focus, and O the other focus of the metal, in which the object is placed.

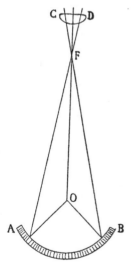

But to return from this digression, I told you that light is not similar or homogeneal, but consists of difform rays, some of which are more refrangible than others; so that of those which are alike incident on the same medium some shall be more refracted than others, and that not by any virtue of the glass or other external cause, but from a predisposition which every particular ray has to suffer a particular degree of refraction.

I shall now proceed to acquaint you with another more notable difformity in its rays, wherein the *origin of colors* is unfolded, concerning which I shall lay down the *doctrine* first and then, for its examination, give you an instance or two of the *experiments,* as a specimen of the rest.

The doctrine you will find comprehended and illustrated in the following propositions:

1. As the rays of light differ in degrees of refrangibility, so they also differ in their disposition to exhibit this or that particular color. Colors are not *qualifications of light,* derived from refractions or reflections of natural bodies (as it is generally believed), but *original* and *connate properties,* which in divers rays are divers. Some rays are disposed to exhibit a red color and no other, some a yellow and no other, some a green and no other, and so of the rest. Nor are there only rays proper and particular to the more eminent colors, but even to all their intermediate gradations.

2. To the same degree of refrangibility ever belongs the same color, and to the same color ever belongs the same degree of refrangibility. The least refrangible rays are all disposed to exhibit a red color, and contrarily those rays which are disposed to exhibit a red color are all the least refrangible; so the most refrangible rays are all disposed to exhibit a deep violet color, and contrarily those which are apt to exhibit such a violet color are all the most refrangible. And so to all the intermediate colors in a continued series belong intermediate degrees of refrangibility. And this analogy between colors and refrangibility is very precise and strict, the rays always either exactly agreeing in both or proportionally disagreeing in both.

3. The species of color and degree of refrangibility proper to any particular sort of rays is not mutable by refraction, nor by reflection from natural bodies, nor by any other cause that I could yet observe. When any one sort of rays has been well parted

from those of other kinds, it has afterward obstinately retained its color, notwithstanding my utmost endeavors to change it. I have refracted it with prisms and reflected it with bodies which in daylight were of other colors; I have intercepted it with the colored film of air interceding two compressed plates of glass, transmitted it through colored mediums and through mediums irradiated with other sorts of rays, and diversely terminated it; and yet could never produce any new color out of it. It would by contracting or dilating become more brisk or faint, and by the loss of many rays in some cases very obscure and dark, but I could never see it changed *in specie*.

4. Yet seeming transmutations of colors may be made where there is any mixture of divers sorts of rays. For in such mixtures the component colors appear not, but, by their mutual allaying each other, constitute a middling color. And therefore, if by refraction or any other of the aforesaid causes, the difform rays latent in such a mixture be separated, there shall emerge colors different from the color of the composition. Which colors are not newly generated, but only made apparent by being parted; for if they be again entirely mixed and blended together, they will again compose that color which they did before separation. And for the same reason, transmutations made by the convening of divers colors are not real; for when the difform rays are again severed, they will exhibit the very same colors which they did before they entered the composition; as you see, blue and yellow powders, when finely mixed, appear to the naked eye green, and yet the colors of the component corpuscles are not thereby really transmuted, but only blended. For, when viewed with a good microscope, they still appear blue and yellow interspersedly.

5. There are therefore two sorts of colors. The one original and simple, the other compounded of these. The original or primary colors are red, yellow, green, blue, and a violet-purple, together with orange, indigo, and an indefinite variety of intermediate gradations.

6. The same colors *in specie* with these primary ones may be also produced by composition; for a mixture of yellow and blue makes green; of red and yellow makes orange; of orange and

yellowish-green makes yellow. And in general, if any two colors be mixed which in the series of those generated by the prism are not too far distant one from another, they by their mutual alloy compound that color which in the said series appears in the midway between them. But those which are situated at too great a distance do not do so. Orange and indigo produce not the intermediate green, nor scarlet and green the intermediate yellow.

7. But the most surprising and wonderful composition was that of *whiteness*. There is no one sort of rays which alone can exhibit this. It is ever compounded, and to its composition are requisite all the aforesaid primary colors, mixed in a due proportion. I have often with admiration beheld that, all the colors of the prism being made to converge and thereby to be again mixed as they were in the light before it was incident upon the prism, reproduced light entirely and perfectly white, and not at all sensibly differing from a direct light of the sun, unless when the glasses I used were not sufficiently clear, for then they would a little incline it to their color.

8. Hence therefore it comes to pass that whiteness is the usual color of light; for light is a confused aggregate of rays indued with all sorts of colors, as they are promiscuously darted from the various parts of luminous bodies. And of such a confused aggregate, as I said, is generated whiteness, if there be a due proportion of the ingredients; but if any one predominate, the light must incline to that color, as it happens in the blue flame of brimstone, the yellow flame of a candle, and the various colors of the fixed stars.

9. These things considered, the manner how colors are produced by the prism is evident. For of the rays constituting the incident light, since those which differ in color proportionally differ in refrangibility, they by their unequal refractions must be severed and dispersed into an oblong form in an orderly succession, from the least refracted scarlet to the most refracted violet. And for the same reason it is that objects, when looked upon through a prism, appear colored. For the difform rays, by their unequal refractions, are made to diverge toward several parts of the retina, and there express the images of things colored, as in the former case they

did the sun's image upon a wall. And by this inequality of refractions they become, not only colored, but also very confused and indistinct.

10. Why the colors of the rainbow appear in falling drops of rain is also from hence evident. For those drops which refract the rays disposed to appear purple in greatest quantity to the spectator's eye refract the rays of other sorts so much less as to make them pass beside it; and such are the drops on the inside of the primary Bow and on the outside of the secondary or exterior one. So those drops which refract in greatest plenty the rays which are apt to appear red toward the spectator's eye refract those of other sorts so much more as to make them pass beside it; and such are the drops on the exterior part of the primary and interior part of the secondary Bow.

11. The odd phenomena of an infusion of *lignum nephriticum,* leaf gold, fragments of colored glass, and some other transparently colored bodies, appearing in one position of one color and of another in another, are on these grounds no longer riddles. For those are substances apt to reflect one sort of light and transmit another, as may be seen in a dark room by illuminating them with similar or uncompounded light. For then they appear of that color only with which they are illuminated, but yet in one position more vivid and luminous than in another, accordingly as they are disposed more or less to reflect or transmit the incident color.

12. From hence also is manifest the reason of an unexpected experiment which Mr. Hooke somewhere in his *Micrography* relates to have made with two wedgelike transparent vessels, filled the one with a red, the other with a blue liquor, namely, that though they were severally transparent enough, yet both together became opaque; for if one transmitted only red and the other only blue, no rays could pass through both.

13. I might add more instances of this nature, but I shall conclude with this general one, that the colors of all natural bodies have no other origin than this that they are variously qualified to reflect one sort of light in greater plenty than another. And this I have experimented in a dark room by illuminating those bodies with uncompounded light of divers colors. For by that means any

body may be made to appear of any color. They have there no appropriate color, but ever appear of the color of the light cast upon them; but yet with this difference that they are most brisk and vivid in the light of their own daylight color. Minium [c] appears there of any color indifferently with which it is illustrated, but yet most luminous in red, and so bise [c] appears indifferently of any color with which it is illustrated, but yet most luminous in blue. And therefore minium reflects rays of any color, but most copiously those indued with red; and consequently when illustrated with daylight, that is, with all sorts of rays promiscuously blended, those qualified with red shall abound most in the reflected light and by their prevalence cause it to appear of that color. And for the same reason bise, reflecting blue most copiously, shall appear blue by the excess of those rays in its reflected light; and the like of other bodies. And that this is the entire and adequate cause of their colors is manifest because they have no power to change or alter the colors of any sort of rays incident apart, but put on all colors indifferently with which they are enlightened.

These things being so, it can be no longer disputed whether there be colors in the dark, nor whether they be the qualities of the object we see, nor perhaps whether light be a body. For since colors are the *qualities* of light, having its rays for their entire and immediate subject, how can we think those rays qualities also unless one quality may be the subject of and sustain another, which in effect is to call it *substance*. We should not know bodies for substances were it not for their sensible qualities, and the principal of those being now found due to something else, we have as good reason to believe that to be a substance also.

Besides, whoever thought any quality to be a *heterogenous* aggregate, such as light is discovered to be[?] But to determine more absolutely what light is, after what manner refracted, and by what modes or actions it produces in our minds the phantasms of colors, is not so easy. And I shall not mingle conjectures with certainties.

Reviewing what I have written, I see the discourse itself will

---

[c] [*Minium*: an earthy red color; red lead. *Bise*: red and blue powders or pigment; "a light blue color prepared from smalt," *New English Dictionary*.]

lead to divers experiments sufficient for its examination, and therefore I shall not trouble you further than to describe one of those which I have already insinuated.

In a darkened room, make a hole in the shut of a window, whose diameter may conveniently be about a third part of an inch, to admit a convenient quantity of the sun's light; and there place a clear and colorless prism to refract the entering light toward the further part of the room, which, as I said, will thereby be diffused into an oblong colored image. Then place a lens of about three-foot radius (suppose a broad object glass of a three-foot telescope) at the distance of about four or five feet from thence, through which all those colors may at once be transmitted and made by its refraction to convene at a further distance of about ten or twelve feet. If at that distance you intercept this light with a sheet of white paper, you will see the colors converted into whiteness again by being mingled. But it is requisite that the prism and lens be placed steady, and that the paper on which the colors are cast be moved to and fro; for by such motion you will not only find at what distance the whiteness is most perfect, but also see how the colors gradually convene and vanish into whiteness, and afterward having crossed one another in that place where they compound whiteness are again dissipated and severed, and in an inverted order retain the same colors which they had before they entered the composition. You may also see that, if any of the colors at the lens be intercepted, the whiteness will be changed into the other colors. And therefore, that the composition of whiteness be perfect, care must be taken that none of the colors fall beside the lens. In the annexed design of this experiment, ABC expresses the prism set endwise to sight, close by the hole F of the window EG.

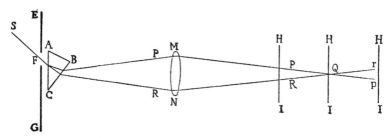

Its vertical angle ACB may conveniently be about sixty degrees; MN designates the lens. Its breadth two and a half or three inches. SF one of the straight lines in which difform rays may be conceived to flow successively from the sun. FP and FR two of those rays unequally refracted which the lens makes to converge toward Q and, after decussation, to diverge again. And HI the paper, at divers distances, on which the colors are projected, which in Q constitutes whiteness but are red and yellow in R and r, and blue and purple in P and p.

If you proceed further to try the impossibility of changing any uncompounded color (which I have asserted in the third and thirteenth Propositions), it is requisite that the room be made very dark, lest any scattering light, mixing with the color, disturb and alloy it, and render it compound, contrary to the design of the experiment. It is also requisite that there be a more perfect separation of the colors than, after the manner above described, can be made by the refraction of one single prism; and how to make such further separations will scarce be difficult to them that consider the discovered laws of refractions. But if trial shall be made with colors not thoroughly separated, there must be allowed changes proportional to the mixture. Thus if compound yellow light fall upon blue bise, the bise will not appear perfectly yellow but rather green; because there are in the yellow mixture many rays indued with green, and green being less remote from the usual blue color of bise than yellow is the more copiously reflected by it.

In like manner, if any one of the prismatic colors, suppose red, be intercepted on design to try the asserted impossibility of reproducing that color out of the others which are pretermitted, it is necessary either that the colors be very well parted before the red be intercepted or that, together with the red, the neighboring colors into which any red is secretly dispersed (that is, the yellow, and perhaps green too) be intercepted, or else that allowance be made for the emerging of so much red out of the yellow green as may possibly have been diffused and scatteringly blended in those colors. And if these things be observed, the new production of red or any intercepted color will be found impossible.

This, I conceive, is enough for an introduction to experiments of this kind, which if any of the Royal Society shall be so curious as to prosecute I should be very glad to be informed with what success, that, if anything seem to be defective or to thwart this relation, I may have an opportunity of giving further direction about it or of acknowledging my errors, if I have committed any.

2. ON THE SCIENCE OF COLORS

From a letter to Oldenburg [d]

. . . I said, indeed, that the science of colors was mathematical and as certain as any other part of optics; but who knows not that optics, and many other mathematical sciences, depend as well on mathematical demonstration? And the absolute certainty of a science cannot exceed the certainty of its principles. Now the evidence by which I asserted the propositions of colors is in the next words expressed to be from experiments, and so but physical, whence the propositions themselves can be esteemed no more than physical principles of a science. And if these principles be such that on them a mathematician may determine all the phenomena of colors that can be caused by refractions and that, by disputing or demonstrating after what manner and how much, those refractions do separate or mingle the rays in which several colors are originally inherent, I suppose the science of colors will be granted mathematical and as certain as any part of optics. And this may be done, I have good reason to believe, because ever since I became first acquainted with these principles I have, with constant success in the events, made use of them for this purpose.

[d] [Cambridge, July 11, 1672. *Opera Omnia* IV, p. 342. These interesting remarks on the science of colors are taken from a long reply of Newton's to criticism which Hooke had made of his studies on the nature of light and color.]

3. HYPOTHESIS TOUCHING ON THE THEORY OF LIGHT AND COLORS [9]

From a letter to Oldenburg [e]

Sir,

In my answer to Mr. Hooke, you may remember I had occasion to say something of hypotheses, where I gave a reason why all allowable hypotheses in their genuine constitution should be conformable to my theories, and said of Mr. Hooke's hypothesis that I took the most free and natural application of it to phenomena to be this: "That the agitated parts of bodies, according to their several sizes, figure, and motions, do excite vibrations in the ether of various depths or bignesses, which, being promiscuously propagated through that medium to our eyes, effect in us a sensation of light of a white color; but if by any means those of unequal bignesses be separated from one another, the largest beget a sensation of a red color, the least or shortest of a deep violet, and the intermediate ones of intermediate colors, much after the manner that bodies, according to their several sizes, shapes, and motions, excite vibrations in the air of various bignesses, which, according to those bignesses, make several tones in sound, etc." I was glad to understand, as I apprehended from Mr. Hooke's discourse at my last being at one of your assemblies, that he had changed his former notion of all colors being compounded of only two original ones, made by the two sides of an oblique pulse, and accommodated his hypothesis to this my suggestion of colors, like sounds, being various, according to the various bigness of the pulses. For this I take to be a more plausible hypothesis than any other described by former authors, because I see not how the colors of thin transparent plates or skins can be handsomely explained without having recourse to etherial pulses. But yet I like another hypothesis bet-

e ["An Hypothesis Explaining the Properties of Light Discoursed of in My Several Papers," in a letter to Oldenburg, January 25, 1675/6. Communicated to the Royal Society, December 9, 1675. Quoted in Brewster, *op. cit.*, Vol. I, pp. 390-409.]

ter, which I had occasion to hint something of in the same letter in these words: "The hypothesis of light's being a body, had I propounded it, has a much greater affinity with the objector's own hypothesis than he seems to be aware of, the vibrations of the ether being as useful and necessary in this as in his. For assuming the rays of light to be small bodies emitted every way from shining substances, those, when they impinge on any retracting or reflecting superficies, must as necessarily excite vibrations in the ether as stones do in water when thrown into it. And supposing these vibrations to be of several depths or thicknesses, accordingly as they are excited by the said corpuscular rays of various sizes and velocities, of what use they will be for explicating the manner of reflection and refraction, the production of heat by the sunbeams, the emission of light from burning, putrifying, or other substances whose parts are vehemently agitated, the phenomena of thin transparent plates and bubbles, and of all natural bodies, the manner of vision, and the difference of colors, as also their harmony and discord, I shall leave to their consideration who may think it worth their endeavor to apply this hypothesis to the solution of phenomena." Were I to assume an hypothesis, it should be this, if propounded more generally so as not to determine what light is, further than that it is something or other capable of exciting vibrations in the ether; for thus it will become so general and comprehensive of other hypotheses as to leave little room for new ones to be invented; and therefore because I have observed the heads of some great virtuosos to run much upon hypotheses, as if my discourses wanted an hypothesis to explain them by, and found that some, when I could not make them take my meaning when I spoke of the nature of light and colors abstractedly, have readily apprehended it when I illustrated my discourse by an hypothesis, for this reason I have here thought fit to send you a description of the circumstances of this hypothesis, as much tending to the illustration of the papers I herewith send you; and though I shall not assume either this or any other hypothesis, not thinking it necessary to concern myself whether the properties of light discovered by me be explained by this or Mr. Hooke's, or any other hypothesis capable of explaining them, yet while I am describing this I shall sometimes, to avoid circumlocution and to represent it more conveniently, speak of it as if I assumed it and propounded it to be believed. This I thought fit to express, that no man may confound this with my other discourses, or measure the certainty of one by the other, or think me obliged to answer objections against this script; for I desire to decline being involved in such troublesome, insignificant disputes.

But to proceed to the hypothesis: 1. It is to be supposed therein that there is an etherial medium, much of the same constitution with air, but far rarer, subtler, and more strongly elastic. Of the existence of this medium, the motion of a pendulum in a glass exhausted of air almost as quickly as in the open air is no inconsiderable argument. But it is not to be supposed that this medium is one uniform matter, but composed partly of the main phlegmatic body of ether, partly of other various etherial spirits, much after the manner that air is compounded of the phlegmatic body of air intermixed with various vapors and exhalations. For the electric and magnetic effluvia and the gravitating principle seem to argue such variety. Perhaps the whole frame of nature may be nothing but various contextures of some certain etherial spirits or vapors, condensed as it were by precipitation, much after the manner that vapors are condensed into water or exhalations into grosser substances, though not so easily condensable; and after condensation wrought into various forms, at first by the immediate hand of the Creator, and ever since by the power of nature, which, by virtue of the command, "increase and multiply," became a complete imitator of the copy set her by the protoplast. Thus perhaps may all things be originated from ether.

At least the electric effluvia seem to instruct us that there is something of an etherial nature condensed in bodies. I have sometimes laid upon a table a round piece of glass about two inches broad, set in a brass ring, so that the glass might be about one eighth or one sixth of an inch from the table, and the air between them inclosed on all sides by the ring, after the manner as if I had whelmed a little sieve upon the table. And then rubbing a pretty while the glass briskly with some rough and raking stuff, till some very little fragments of very thin paper laid on the table under the glass began to be attracted and move nimbly to and fro. After I had done rubbing the glass, the papers would continue a pretty while in various motions, sometimes leaping up to the glass and resting there a while, then leaping down and resting there, then leaping up, and perhaps down and up again, and this sometimes in lines seeming perpendicular to the table, sometimes in oblique ones; sometimes also they would leap up in one arch and down in another divers times together, without sensible resting between; sometimes skip in a bow from one part of the glass to another without touching the table, and sometimes hang by a corner and turn often about very nimbly, as if they had been carried about in the midst of a whirlwind, and be otherwise variously moved, every paper with a divers motion. And upon sliding my finger on the upper side of the glass, though neither the glass nor the enclosed air below were moved thereby, yet would the papers as they hung under the glass receive some new motion, incline this way or that way, accordingly as I moved my finger. Now whence all these irregular motions should spring I cannot imagine, unless from some kind of subtle matter lying condensed in the glass and rarefied by rubbing, as water is rarefied into vapor by heat, and in that rarefaction diffused through the space round the glass to a great distance and made to move and circulate variously, and accordingly to actuate the papers, till it returns into the glass again and be recondensed there. And as this condensed matter by rarefaction into an etherial wind (for by its easy penetrating and circulating through glass I esteem it etherial) may cause these odd motions, and by condensing again may cause electrical attraction with its returning to the glass to succeed in the place of what is there continually recondensed, so may the gravitating attraction of the earth be caused by the continual condensation of some other such-like etherial spirit, not of the main body of phlegmatic ether, but of something very thinly and subtly diffused through it, perhaps of an unctuous or gummy tenacious and springy nature, and bearing much the same relation to ether which the vital aerial spirit requisite for the conservation of flame and vital motions does to air. For if such an etherial spirit may be condensed in fermenting or burning bodies, or otherwise coagulated in the pores of the earth and water into some kind of humid active matter for the continual uses of nature (adhering to the sides of those pores after the manner that vapors condense on the sides of a vessel), the vast body of the earth, which may be everywhere to the very center in perpetual working, may continually condense so much of this spirit as to cause it from above to descend with great celerity for a supply; in which descent it may bear down with it the bodies it pervades with force proportional to the superficies of all their parts it acts upon, nature making a circulation by the slow ascent of as much matter out of the bowels of the earth in an aerial form, which for a time constitutes the atmosphere, but being continually buoyed up by the new air, exhalations, and vapors rising underneath at length (some part of the vapors which return in rain excepted) vanishes again into the etherial spaces, and there perhaps in time relents and is attenuated into its first principle. For nature is a perpetual circulatory worker, generating fluids out of solids, and solids out of fluids; fixed things out of volatile, and volatile out of fixed; subtle out of gross, and gross out of subtle; some things to ascend and make the upper terrestrial juices, rivers, and the atmosphere, and by consequence others to descend for a requital to the former. And as the earth, so perhaps may the sun imbibe this spirit copiously, to conserve his shining and keep the planets from receding further from him; and they that will may also suppose that this spirit affords or carries with it thither the solary fuel and material principle of light, and that the vast etherial spaces between us and the stars are for a sufficient repository for this food of the sun and planets. But this of the constitution of etherial natures by the bye.

In the second place, it is to be supposed that the ether is a vibrating medium like air, only the vibrations far more swift and minute; those of air made by a man's ordinary voice succeeding one another at more than half a foot or a foot distance, but those of ether at a less distance than the hundred-thousandth part of an inch. And as in air the vibrations are some larger than others, but yet all equally swift (for in a ring of bells the sound of every tone is heard at two or three miles' distance in the same order that the bells are struck), so I supposed the etherial vibrations differ in bigness but not in swiftness. Now these vibrations, besides their use in reflection and refraction, may be supposed the chief means by which the parts of fermenting or putrifying substances, fluid liquors, or melted, burning, or other hot bodies, continue in motion, are shaken asunder like a ship by waves and dissipated into vapors, exhalations, or smoke, and light loosed or excited in those bodies, and consequently by which a body becomes a burning coal, and smoke flame; and I suppose flame is nothing but the particles of smoke turned by the access of light and heat to burning coals, little and innumerable.

Thirdly, the air can pervade the bores of small glass pipes, but yet not so easily as if they were wider, and therefore stands at a greater degree of rarity than in the free aerial spaces and at so much greater a degree of rarity as the pipe is smaller, as is known by the rising of water in such pipes to a much greater height than the surface of the stagnating water into which they are dipped. So I suppose ether, though it pervades the pores of crystal, glass, water, and other natural bodies, yet it stands at a greater degree of rarity in those pores than in the free etherial spaces, and at so much a greater degree of rarity as the pores of the body are smaller. Whence it may be that spirit of wine, for instance, though a lighter body, yet having subtler parts and consequently smaller pores than water, is the more strongly refracting liquor. This also may be the principal cause of the cohesion of the parts of solids and fluids, of the springiness of glass and other bodies whose parts slide not one upon another in bending, and of the standing of the mercury in the Torricellian experiment, sometimes to the top of the glass, though a much greater height than twenty-nine inches. For the denser ether which surrounds these bodies must crowd and press their parts together, much after the manner that air surrounding two marbles presses them together if there be little or no air between them. Yea, and that puzzling problem, *by what means the muscles are contracted and dilated to cause animal motion, may receive greater light from hence than from any other means men have hitherto been thinking on.* For if there be any power in man to condense and dilate at will the ether that pervades the muscle, that condensation or dilatation must vary the compression of the muscle made by the ambient ether and cause it to swell or shrink accordingly; for though common water will scarce shrink by compression and swell by relaxation, yet (so far as my observation reaches) spirit of wine and oil will, and Mr. Boyle's experiment of a tadpole shrinking very much by hard compressing the water in which it swam is an argument that animal juices do the same; and as for their various pression by the ambient ether, it is plain that that must be more or less, accordingly as there is more or less ether within to sustain and counterpoise the pressure of that without. If both ethers were equally dense, the muscle would be at liberty as if pressed by neither; if there were no ether within, the ambient would compress it with the whole force of its spring. If the ether within were twice as much dilated as that without, so as to have but half as much springiness, the ambient would have half the force of its springiness counterpoised thereby and exercise but the other half upon the muscle; and so in all other cases the ambient compresses the muscles by the excess of the force of its springiness above that of the springiness of the included. To vary the compression of the muscle, therefore, and so to swell and shrink it, there needs nothing but to change the consistence of the included ether; and a very little change may suffice, if the spring of ether be supposed very strong, as I take it to be many degrees stronger than that of air.

Now for the changing the consistence of the ether, some may be ready to grant that the soul may have an immediate power over the whole ether in any part of the body, to swell or shrink it at will; but then how depends the muscular motion on the nerves? Others, therefore, may be more apt to think it done by some certain etherial spirit included within the *dura mater,* which the soul may have power to contract or dilate at will in any muscle, and so cause it to flow thither through the nerves; but still there is a difficulty why this force of the soul upon it does not take off the power of springiness, whereby it should sustain more or less the force of the outward ether. A third supposition may be that the soul has a power to inspire any muscle with this spirit, by impelling it thither through the nerves; but this too has its difficulties, for it requires a forcible intruding the spring of the ether in the muscles by pressure exerted from the parts of the brain, and it is hard to conceive how so great force can be exercised amidst so tender matter as the brain is; and besides, why does not this etherial spirit, being subtle enough, and urged with so great force, go away through the *dura mater* and skins of the muscle, or at least so much of the other ether go out to make way for this which is crowded in? To take away these difficulties is a digression, but seeing the subject is a deserving one I shall not stick to tell you how I think it may be done.

First, then, I suppose there *is* such a spirit; that is, that the animal spirits are neither like the liquor, vapor, or gas of spirits of wine, but of an etherial nature, subtle enough to pervade the animal juices as freely as the electric or perhaps magnetic effluvia do glass. And to know how the coats of the brain, nerves, and muscles may become a convenient vessel to hold so subtle a spirit, you may consider how liquors and spirits are disposed to pervade, or not pervade, things on other accounts than their subtlety: water and oil pervade wood and stone, which quicksilver does not; and quicksilver, metals which water and oil do not; water and acid spirits pervade salts, which oil and spirit of wine do not; and oil and spirit of wine pervade sulphur, which water and acid spirits do not. So some fluids (as oil and water), though their parts are in freedom enough to mix with one another, yet by some secret principle of *unsociableness* they keep asunder; and some that are *sociable* may become *unsociable* by adding a third thing to one of them, as water to spirit of wine by dissolving salt of tartar in it. The like *unsociableness* may be in etherial natures, as perhaps between the ethers in the vortices of the sun and planets, and the reason why air stands rarer in the bores of small glass pipes and ether in the pores of bodies may be, not want of subtlety, but *sociableness;* and on this ground, if the etherial vital spirit in a man be very *sociable* to the marrow and juices, and *unsociable* to the coats of the brain, nerves, and muscles, or to anything lodged in the pores of those coats, it may be contained thereby, notwithstanding its sublety, especially if we suppose no great violence done to it to squeeze it out and that it may not be altogether so subtle as the main body of ether, though subtle enough to pervade readily the animal juices, and that, as any of it is spent, it is continually supplied by new spirit from the heart.

In the next place, for knowing how this spirit may be used for animal motion, you may consider how some things unsociable are made sociable by the mediation of a third. Water, which will not dissolve copper, will do it if the copper be melted with sulphur. Aquafortis, which will not pervade gold, will do it by addition of a little sal ammoniac or spirit of salt. Lead will not mix in melting with copper; but if a little tin or antimony be added, they mix readily, and part again of their own accord if the antimony be wasted by throwing saltpeter or otherwise. And so lead melted with silver quickly pervades and liquifies the silver in a much less heat than is required to melt the silver alone; but if they be kept in the test till that little substance that reconciled them be wasted or altered, they part again of their own accord. And in like manner the etherial animal spirit in a man may be a mediator between the common ether and the muscular juices, to make them mix more freely; and so, by sending a little of this spirit into any muscle, though so little as to cause no sensible tension of the muscle by its own force, yet by rendering the juices more sociable to the common external ether, it may cause that ether to pervade the muscle of its own accord in a moment more freely and more copiously than it would otherwise do, and to recede again as freely so soon as this mediator of sociableness is retracted; whence, according to what I said above, will proceed the swelling or shrinking of the muscle, and consequently the animal motion depending thereon.

Thus may therefore the soul, by determining this etherial animal spirit or wind into this or that nerve, perhaps with as much ease as air is moved in open spaces, cause all the motions we see in animals; for the making which motions strong it is not necessary that we should suppose the ether within the muscle very much condensed or rarefied by this means, but only that its spring is so very great that a little alteration of its density shall cause a great alteration in the pressure. And what is said of muscular motion may be applied to the motion of the heart, only with this difference that the spirit is not sent thither as into other muscles, but continually generated there by the fermentation of the juices with which its flesh is replenished, and as it is generated let out by starts into the brain, through some convenient *ductus*, to perform those motions in other muscles by inspiration which it did in the heart by its generation. For I see not why the ferment in the heart may not raise as subtle a spirit out of its juices to cause those motions as rubbing does out of a glass to cause electric attraction, or burning out of fuel to penetrate glass, as Mr. Boyle has shown, and calcine by corrosion metals melted therein.

Hitherto I have been contemplating the nature of ether and etherial substances by their effects and uses, and now I come to join therewith the consideration of light.

In the fourth place, therefore, I suppose light is neither ether nor its vibrating motion, but something of a different kind propagated from lucid bodies. They that will may suppose it an aggregate of various peripatetic qualities. Others may suppose it multitudes of unimaginable small and swift corpuscles of various sizes springing from shining bodies at great distances one after another, but yet without a sensible interval of time, and continually urged forward by a principle of motion, which in the beginning accelerates them, till the resistance of the etherial medium equal the force of that principle, much after the manner that bodies let fall in water are accelerated till the resistance of the water equals the force of gravity. God, who gave animals motion beyond our understanding, is, without doubt, able to implant other principles of motions in bodies which we may understand as little. Some would readily grant this may be a spiritual one; yet a mechanical one might be shown, did not I think it better to pass it by. But they that like not this may suppose light any other corporeal emanation, or an impulse or motion of any other medium or etherial spirit diffused through the main body of ether, or what else they imagine proper for this purpose. To avoid dispute and make this hypothesis general, let every man here take his fancy; only whatever light be, I would suppose it consists of successive rays differing from one another in contingent circumstances, as bigness, force, or vigor, like as the sands on the shore, the waves of the sea, the faces of men, and all other natural things of the same kind differ, it being almost impossible for any sort of things to be found without some contingent variety. And further, I would suppose it diverse from the vibrations of the ether, because (besides that, were it those vibrations, it ought always to verge copiously in crooked lines into the dark or quiescent medium, destroying all shadows, and to comply readily with any crooked pores or passages as sounds do) I see not how any superficies (as the side of a glass prism on which the rays within are incident at an angle of about forty degrees) can be totally opaque. For the vibrations beating against the refracting confine of the rarer and denser ether must needs make that pliant superficies undulate, and those undulations will stir up and propagate vibrations on the other side. And further, how light, incident on very thin skins or plates of any transparent body, should for many successive thicknesses of the plate in arithmetical progression be alternately reflected and transmitted, as I find it is, puzzles me as much. For though the arithmetical progression of those thicknesses, which reflect and transmit the rays alternately, argues that it depends upon the number of vibrations between the two superficies of the plate, whether the ray shall be reflected or transmitted, yet I cannot see how the number should vary the case, be it greater or less, whole or broken, unless light be supposed something else than these vibrations. Something indeed I could fancy toward helping the two last difficulties, but nothing which I see not insufficient.

Fifthly, it is to be supposed that light and ether mutually act upon one another, ether in refracting light and light in warming ether, and that the densest ether acts most strongly. When a ray therefore moves through ether of uneven density, I suppose it most pressed, urged, or acted upon by the medium on that side toward the denser ether and receives a continual impulse or ply from that side to recede toward the rarer, and so is accelerated if it move that way or retarded if the contrary. On this ground, if a ray move obliquely through such an unevenly dense medium (that is, obliquely to those imaginary superficies which run through the equally dense parts of the medium and may be called the refracting superficies), it must be incurved, as it is found to be by observation in water, whose lower parts were made gradually more salt, and so more dense than the upper. And this may be the ground of all refraction and reflection. For as the rarer air within a small glass pipe, and the denser without, are not distinguished by a mere mathematical superficies, but have air between them at the orifice of the pipe running through all intermediate degrees of density, so I suppose the refracting superficies of ether between unequally dense mediums to be, not a mathematical one, but of some breadth, the ether therein at the orifices of the pores of the solid body being of all intermediate degrees of density between the rarer and denser etherial mediums; and the refraction I conceive to proceed from the continual incurvation of the ray all the while it is passing the physical superficies. Now if the motion of the ray be supposed in this passage to be increased or diminished in a certain proportion, according to the difference of the densities of the etherial mediums, and the addition or detraction of the motion be reckoned in the perpendicular from the refracting superficies, as it ought to be, the sines of incidence and refraction will be proportional according to what Descartes has demonstrated.

The ray, therefore, in passing out of the rarer medium into the denser, inclines continually more and more toward parallelism with the refracting superficies; and if the differing densities of the mediums be not so great, nor the incidence of the ray so oblique as to make it parallel to that superficies before it gets through, then it goes through and is refracted; but if through the aforesaid causes the ray becomes parallel to that superficies before it can get through, then it must turn back and be reflected....

... This may be the cause and manner of reflection, when light tends from the rarer toward the denser ether; but to know how it should be reflected when it tends from the denser toward the rarer you are further to consider how fluids near their superficies are less pliant and yielding than in their more inward parts, and if formed into thin plates or shells they become much more stiff and tenacious than otherwise. Thus things which readily fall in water, if let fall upon a bubble of water, they do not easily break through it, but are apt to slide down by the sides of it, if they are not too big and heavy. So if two well-polished convex glasses, ground on very large spheres, be laid one upon the other, the air between them easily recedes till they almost touch, but then begins to resist so much that the weight of the upper glass is too little to bring them together, so as to make the black (mentioned in the other papers I send you) appear in the midst of the rings of colors. And if the glasses be plain, though no broader than a twopence, a man with his whole strength is not able to press all the air out from between them, so as to make them fully touch. You may observe also that insects will walk upon water without wetting their feet, and the water bearing them up; also motes falling upon water will often lie long upon it without being wetted. And so I suppose ether in the confine of two mediums is less pliant and yielding than in other places, and so much the less pliant by how much the mediums differ more in density; so that in passing out of denser ether into rarer, when there remains but a very little of the denser ether to be passed through, a ray finds more than ordinary difficulty to get through, and so great difficulty where the mediums are of a very differing density as to be reflected by incurvation after the manner described above, the parts of ether on the side where they are less pliant and yielding acting upon the ray much after the manner that they would do were they denser there than on the other side; for the resistance of the medium ought to have the same effect on the ray from whatsoever cause it arises. And this I suppose may be the cause of the reflection of quicksilver and other metalline bodies. It must also concur to increase the reflective virtue of the superficies when rays tend out of the rarer medium into the denser, and in that case therefore, the reflection having a double cause, ought to be stronger than in the ether, as it is apparently. But in refraction this rigid tenacity or unpliableness of the superficies need not be considered, because so much as the ray is thereby bent in passing to the most tenacious and rigid part of the superficies, so much is it thereby unbent again in passing on from thence through the next parts gradually less tenacious.

Thus may rays be refracted by some superficies and reflected by others, be the medium they tend into denser or rarer. But it remains further to be explained how rays alike incident on the same superficies (suppose of crystal, glass, or water) may be, at the same time, some refracted, others reflected; and for explaining this, I suppose that the rays, when they impinge on the rigid resisting etherial superficies, as they are acted upon by it so they react upon it and cause vibrations in it, as stones thrown into water do in its surface; and that these vibrations are propagated every way into both the rarer and denser mediums, as the vibrations of air which cause sound are from a stroke, but yet continue strongest where they began, and alternately contract and dilate the ether in that physical superficies. For it is plain by the heat which light produces in bodies that it is able to put their parts in motion, and much more to heat and put in motion the more tender ether; and it is more probable that it communicates motion to the gross parts of bodies by the mediation of ether than immediately, as, for instance, in the inward parts of quicksilver, tin, silver, and other very opaque bodies, by generating vibrations that run through them than by striking the outward parts only without entering the body. The shock of every single ray may generate many thousand vibrations and, by sending them all over the body, move all the parts, and that perhaps with more motion than it could move one single part by an immediate stroke; for the vibrations, by shaking each particle backward and forward, may every time increase its motion, as a ringer does a bell by often pulling it, and so at length move the particles to a very great degree of agitation, which neither the simple shock of a ray nor any other motion in the ether besides a vibrating one could do. Thus in air shut up in a vessel the motion of its parts caused by heat, howsoever violent, is unable to move the bodies hung in it with either a trembling or progressive motion; but if air be put into a vibrating motion by beating a drum or two, it shakes glass windows, the whole body of a man, and other massy things, especially those of a congruous tone; yea, I have observed it manifestly shake under my feet a cellared freestone floor of a large hall, so as I believe the immediate stroke of five hundred drumsticks could not have done, unless perhaps quickly succeeding one another at equal intervals of time. Etherial vibrations are therefore the best means by which such a subtle agent as light can shake the gross particles of solid bodies to heat them. And so supposing that light impinging on a refracting or reflecting etherial superficies puts it into a vibrating motion, that physical superficies being by the perpetual appulse of rays always kept in a vibrating motion and the ether therein continually expanded and compressed by turns, if a ray of light impinge upon it while it is much compressed I suppose it is then too dense and stiff to let the ray pass through, and so reflects it; but the rays that impinge on it at other times, when it is either expanded by the interval of two vibrations or not too much compressed and condensed, go through and are refracted.

These may be the causes of refractions and reflections in all cases, but for understanding how they come to be so regular, it is further to be considered that, as in a heap of sand, although the surface be rugged, yet if water be poured on it to fill its pores, the water, so soon as its pores are filled, will evenly overspread the surface, and so much the more evenly as the sand is finer; so, although the surface of all bodies, even the most polished, be rugged, as I conceive, yet when that ruggedness is not too gross and coarse, the refracting etherial superficies may evenly overspread it. In polishing glass or metal, it is not to be imagined that sand, putty, or other fretting powders should wear the surface so regularly as to make the front of every particle exactly plane, and all those planes look the same way, as they ought to do in well-polished bodies, were reflection performed by their parts; but that those fretting powders should wear the bodies first to a coarse ruggedness, such as is sensible, and then to a finer and finer ruggedness, till it be so fine that the etherial superficies evenly overspreads it, and so makes the body put on the appearance of a polish, is a very natural and intelligible supposition. So in fluids it is not well to be conceived that the surfaces of their parts should be all plain and the planes of the superficial parts always kept looking all the same way, notwithstanding that they are in perpetual motion; and yet without these two suppositions, the superficies of fluids could not be so regularly reflective as they are, were the reflection done by the parts themselves and not by an etherial superficies evenly overspreading the fluid.

Further, concerning the regular motion of light, it might be suspected whether the various vibrations of the fluid through which it passes may not much disturb it; but that suspicion I suppose will vanish by considering that if at any time the foremost part of an oblique wave begin to turn it awry, the hindermost part by a contrary action must soon set it straight again.

Lastly, because without doubt there are in every transparent body pores of various sizes, and I said that ether stands at the greatest rarity in the smallest pores, hence the ether in every pore should be of a differing rarity, and so light be refracted in its passage out of every pore into the next, which would cause a great confusion and spoil the body's transparency; but considering that the ether in all dense bodies is agitated by continual vibrations, and these vibrations cannot be performed without forcing the parts of ether forward and backward from one pore to another by a kind of tremor, so that the ether which one moment is in a great pore is the next moment forced into a less, and on the contrary this must evenly spread the ether into all the pores not exceeding some certain bigness, suppose the breadth of a vibration, and so make it of an even density throughout the transparent body, agreeable to the middle sort of pores. But where the pores exceed a certain bigness, I suppose the ether suits its density to the bigness of the pore or to the medium within it, and so being of a divers density from the ether that surrounds it refracts or reflects light in its superficies, and so makes the body where many such interstices are appear opaque.

Thus much of refraction, reflection, transparency, and opacity, and now to explain colors. I suppose that as bodies of various sizes, densities, or tensions do by percussion or other action excite sounds of various tones, and consequently vibrations in the air of various bignesses, so when the rays of light, by impinging on the stiff refracting superficies, excite vibrations in the ether, those rays, whatever they be, as they happen to differ in magnitude, strength, or vigor, excite vibrations of various bignesses—the biggest, strongest, or most potent rays, the largest vibrations; and others shorter according to their bigness, strength, or power—and therefore the ends of the capillamenta of the optic nerve, which front or face the retina, being such refracting superficies, when the rays impinge upon them, they must there excite these vibrations, which vibrations (like those of sound in a trumpet) will run along the aqueous pores or crystalline pith of the capillamenta, through the optic nerves into the sensorium (which light itself cannot do), and there, I suppose, affect the sense with various colors according to their bigness and mixture—the bigness with the strongest colors, reds and yellows; the least with the weakest, blues and violets; the middle with green; and a confusion of all with white—much after the manner that in the sense of hearing nature makes use of aerial vibrations of several bignesses to generate sounds of divers tones, for the analogy of nature is to be observed. And further, as the harmony and discord of sounds proceed from the proportions of the aerial vibrations, so may the harmony of some colors, as of a golden and blue, and the discord of others, as of red and blue, proceed from the proportions of the etherial. And possibly color may be distinguished into its principal degrees—red, orange, yellow, green, blue, indigo, and deep violet—on the same ground that sound within an eighth is graduated into tones. For some years past the prismatic colors, being in a well-darkened room, cast perpendicularly upon a paper about two-and-twenty foot distant from the prism, I desired a friend to draw with a pencil lines across the image or pillar of colors where every one of the seven aforenamed colors was most full and brisk, and also where he judged the truest confines of them to be, whilst I held the paper so that the said image might fall within a certain compass marked on it. And this I did, partly because my own eyes are not very critical in distinguishing colors, partly because another to whom I had not communicated my thoughts about this matter could have nothing but his eyes to determine his fancy in making those marks. This observation we repeated divers times, both in the same and divers days, to see how the marks on several papers would agree; and comparing the observations, though the just confines of the colors are hard to be assigned because they passed into one another by insensible gradation, yet the differences of the observations were but little, especially toward the red end, and taking means between those differences that were, the length of the image (reckoned not by the distance of the verges of the semicircular ends, but by the distance of the centers of those semicircular ends, or length of the straight sides, as it ought to be) was divided in about the same proportion that a string is between the end and the middle to sound the tones in an eighth . . . .

Now for the cause of these and suchlike colors made by refraction, the biggest or strongest rays must penetrate the refracting superficies more freely and easily than the weaker, and so be less turned awry by it, that is, less refracted; which is as much as to say, the rays which make red are least refrangible, those which make blue or violet most refrangible, and others otherwise refrangible according to their color. Whence if the rays which come promiscuously from the sun be refracted by a prism, as in the aforesaid experiment, those of several sorts being variously refracted must go to several places on an opposite paper or wall, and so parted exhibit every one their own colors, which they could not do while blended together. And because refraction only severs them and changes not the bigness of strength of the ray, thence it is that, after they are once well-severed, refraction cannot make any further changes in their color. On this ground may all the phenomena of refractions be understood.

4. PERCEPTION[10]

From the Optics [t]

. . . When a man views any object . . . the light which comes from the several points of the object is refracted by the transparent skins and humors of the eye (that is, by the outward coat . . . called the *tunica cornea,* and by the crystalline humor . . . which is beyond the pupil . . .) as to converge and meet again in so many points in the bottom of the eye, and there to paint the picture of the object upon the skin (called the *tunica retina*) with which the bottom of the eye is covered . . . and these pictures, propagated by motion along the fibers of the optic nerves in the brain, are the cause of vision. For accordingly as these pictures are perfect or imperfect, the object is seen perfectly or imperfectly. . . .

[t] [*Opticks: or, a Treatise of the Reflections, Refractions, Inflections, and Colours of Light* (3rd ed., corrected, London, 1721), p. 12.]

DEFINITION [g]

The homogeneal light and rays which appear red, or rather make objects appear so, I call rubrific, or red-making; those which make objects appear yellow, green, blue, and violet I call yellow-making, green-making, blue-making, violet-making, and so of the rest. And if at any time I speak of light and rays as colored or endued with colors, I would be understood to speak, not philosophically and properly, but grossly and accordingly to such conceptions as vulgar people in seeing all these experiments would be apt to frame. For the rays, to speak properly, are not colored. In them there is nothing else than a certain power and disposition to stir up a sensation of this or that color. For as sound in a bell or musical string, or other sounding body, is nothing but a trembling motion, and in the air nothing but that motion propagated from the object, and in the sensorium it is a sense of that motion under the form of sound, so colors in the object are nothing but a disposition to reflect this or that sort of rays more copiously than the rest; in the rays they are nothing but their dispositions to propagate this or that motion into the sensorium, and in the sensorium they are sensations of those motions under the forms of colors.

For his Hon^d friend D^r. W^m. Briggs [h]

S^r

Though I am of all men grown the most shy of setting pen to paper about anything that may lead into disputes, yet your friendship overcomes me so far as that I shall set down my suspicions about your theory; yet on this condition that, if I can write but plain enough to make you understand me, I may leave all to your use without pressing it further on. For I design not to confute or convince you, but only to present and submit my thoughts to your consideration and judgment.

... It may be further considered that the cause of an object's appearing one to both eyes is not its appearing of the same color,

---

[g] [*Ibid.*, pp. 108-9.]
[h] [From a letter to William Briggs, quoted in Edleston, *op. cit.*, p. 265.]

form, and bigness to both, but in the same situation or place. Distort one eye, and you will see the coincident images of the object divide from one another and one of them remove from the other upward, downward, or sideways to a greater or less distance, according as the distortion is; and when the eyes are let return to their natural posture, the two images advance toward one another till they become coincident and by that coincidence appear but one. If we would then know why they appear but one, we must inquire why they appear in one and the same place; and if we would know the cause of that, we must inquire why in other cases they appear in divers places variously situated and distant one from another. For that which can make their distance greater or less can make it none at all. Consider what is the cause of their being in the same altitude when one is directly to the right hand, the other to the left, and what of their being in the same coast or point of the compass when one is directly over the other; these two causes joined will make them in the same altitude and coast at once that is in the same place. The cause of situations is therefore to be inquired into. Now, for finding out this, the analogy will stand between the situations of sound and the situations of visible things, if we will compare these two senses. But the situations of sounds depend not on their tones. I can judge from whence an echo or other sound comes though I see not the sounding body, and this judgment depends not at all on the tone. I judge it not from east because acute, from west because grave; but be the tone what it will, I judge it from hence or thence by some other principle. And by that principle I am apt to think a blind man may distinguish unisons one from another when the one is on his right hand the other on his left. And were our ears as good and accurate at distinguishing the coasts of audibles as our eyes are at distinguishing the coasts of visibles, I conceive we should judge no two sounds the same for being unisons unless they came so exactly from the same coast as not to vary from one another a sensible point in situation to any side. Suppose, then, you had to do with one of so accurate an ear in distinguishing the situation of sounds. How would you deal with him? Would you tell him that you heard all unisons as but one sound? He would tell you he had a better ear than so. He accounted no sounds the same which differed in situation; and if your eyes were no better at the situation of things than your ears, you would perhaps think all objects the same which were of the same color. But for his part he found that the like tension of strings and other sounding bodies did not make sounds one, but only of the same tone; and therefore, not allowing the supposition that it does make them one, the inference from thence that the like tension of optic fibers made the object to the two eyes appear one, he did not think himself obliged to admit. As he found that tones depended on those tensions so perhaps might colors, but the situation of audibles depended not on those tensions; and therefore if the two senses hold analogy with one another, that of visibles does not, and consequently the union of visibles as well as audibles, which depends on the agreement of situation as well as of color or tone, must have some other cause.

But to leave this imaginary disputant, let us now consider what may be the cause of the various situations of things to the eyes. If when we look but with one eye it be asked why objects appear thus and thus situated one to another, the answer would be because they are really so situated among themselves and make their colored pictures in the retina so situated one to another as they are; and those pictures transmit motional pictures into the sensorium in the same situation, and by the situation of those motional pictures one to another the soul judges of the situation of things without. In like manner when we look with two eyes distorted so as to see the same object double, if it be asked why those objects appear in this or that situation and distance one from another, the answer should be because through the two eyes are transmitted into the sensorium two motional pictures by whose situation and distance then from one another the soul judges she sees two things so situate and distant. And if this be true, then the reason why, when the distortion ceases and the eyes return to their natural posture, the doubled object grows a single one is that the two motional pictures in the sensorium come together and become coincident.

But, you will say, how is this coincidence made? I answer, what if I know not? Perhaps in the sensorium, after some such way as the Cartesians would have believed or by some other way. Perhaps by the mixing of the marrow of the nerves in their juncture before they enter the brain, the fibers on the right side of each eye going to the right side of the head, those on the left side to the left. If you mention the experiment of the nerve shrunk all the way on one side of the head, that might be either by some unkind juice abounding more on one side the head than on the other, or by the shrinking of the coat of the nerve whose fibers and vessels for nourishment perhaps do not cross in the juncture as the fibers of the marrow may do. And it is more probable yet the stubborn coat, being vitiated or wanting due nourishment, shrank and made the tender marrow yield to its capacity than that the tender marrow, by shrinking, should make the coat yield. I know not whether you would have the *succus nutricius* run along the marrow. If you would, it is an opinion not yet proved, and so not fit to ground an argument on. If you say, yet in the chameleon and fishes the nerves only touch one another without mixture and sometimes do not so much as touch, it is true, but makes altogether against you. Fishes look one way with one eye, the other way with the other; the chameleon looks up with one eye, down with the other, to the right hand with this, to the left with that, twisting his eyes severally this way or that way as he pleases. And in these animals, which do not look the same way with both eyes, what wonder if the nerves do not join? To make them join would have been to no purpose, and nature does nothing in vain. But then whilst in these animals where it is not necessary they are not joined, in all others which look the same way with both eyes, so far as I can yet learn, they are joined. Consider, therefore, for what reason they are joined in the one and not in the other. For God, in the frame of animals, has done nothing without reason.

. . . You have now the sum of what I can think of worth objecting set down in a tumultuary way, as I could get time from my Sturbridge Fair friends. If I have anywhere expressed myself in a more peremptory way than becomes the weakness of the argument, pray look on that as done, not in earnestness, but for the mode of discoursing. Whether anything be so material as that it may prove anyway useful to you I cannot tell. But pray accept of it as written for that end. For having laid philosophical speculations aside, nothing but the gratification of a friend would easily invite me to so large a scribble about things of this nature.

S^r I am

Y^{or} humble servant

Is. NEWTON

Trinity College, Cambridge, September 12, 1682

From a Manuscript ^{i}

...Light seldom strikes upon the parts of gross bodies (as may be seen in its passing through them); its reflection and refraction is made by the diversity of ethers, and therefore its effect upon the retina can only be to make this vibrate, which motion then must be either carried in the optic nerves to the sensorium or produce other motions that are carried thither. Not the latter, for water is too gross for such subtle impressions; and as for animal spirits, though I tied a piece of the optic nerve at one end and warmed it in the middle, to see if any airy substance by that means would disclose itself in bubbles at the other end, I could not spy the least bubble; a little moisture only, and the marrow itself squeezed out. And indeed they that know how difficultly air enters small pores of bodies have reason to suspect that an airy body, though much finer than air, can pervade, and without violence (as it ought to do), the small pores of the brain and nerves, I should say of water because those pores are filled with water; and if it could it would be too subtle to be imprisoned by the *dura mater* and skull, and might pass for ether. However, what need of such spirits? Much motion is ever lost by communication, especially betwixt bodies of different constitutions. And therefore it can no way be conveyed to the sensorium so entirely as by the ether itself. Nay, granting me but that there are pipes filled with a pure transparent liquor passing from the eye to the sensorium and the vibrating motion of the ether will of necessity run along thither. For nothing interrupts that motion but reflecting surfaces; and therefore also that motion cannot stray through the reflecting surfaces of the pipe, but must run along (like a sound in a trunk) entire to the sensorium. And

---

^{i} [Quoted in Brewster, *op. cit.*, Vol. I, pp. 435-6.]

that vision thus made is very comfortable to the sense of hearing, which is made by like vibrations.

5. ON GRAVITY [11]

PROPOSITION LXXVI [j]

If spheres be however dissimilar (as to density of matter and attractive force) in the same ratio onward from the center to the circumference, but everywhere similar at every given distance from the center, on all sides round about; and the attractive force of every point decreases as the square of the distance of the body attracted: I say that the whole force with which one of these spheres attracts the other will be inversely proportional to the square of the distance of the centers.

COROLLARY III

The motive attractions, or the weights of the spheres toward one another, will be at equal distances of the centers conjointly as the attracting and attracted spheres; that is, as the products arising from multiplying the spheres into each other.

COROLLARY IV

And at unequal distances directly as those products and inversely as the squares of the distances between the centers.

The intensities of the forces and the resulting motions in individual cases [k]

Therefore the absolute force of every globe is as the quantity of matter which the globe contains; but the motive force by which

[j] [Proposition LXXVI and Corollaries III and IV, *Principia*, Bk. I.]

[k] [*System of the World*, Sec. 26. Section 25 had concluded with the statement that ". . . two such globes will (by Proposition LXXVI) attract one the other with a force decreasing inversely as the square of the distance between their centers."]

every globe is attracted toward another and which, in terrestrial bodies, we commonly call their weight, is as the content under the quantities of matter in both globes divided by the square of the distance between their centers (by Corollary IV, Proposition LXXVI), to which force the quantity of motion by which each globe in a given time will be carried toward the other is proportional. And the accelerative force by which every globe according to its quantity of matter is attracted toward another is as the quantity of matter in that other globe divided by the square of the distance between the centers of the two [1] ... to which force the velocity by which the attracted globe will, in a given time, be carried toward the other is proportional. And from these principles well understood, it will now be easy to determine the motions of the celestial bodies among themselves.

PROPOSITION IV [m]

That the moon gravitates toward the earth, and by the force of gravity is continually drawn off from a rectilinear motion and retained in its orbit.

SCHOLIUM

The demonstration of this Proposition may be more diffusely explained after the following manner. Suppose several moons to revolve about the earth, as in the system of Jupiter or Saturn; the periodic times of these moons (by the argument of induction) would observe the same law which Kepler found to obtain among the planets, and therefore their centripetal forces would be inversely as the squares of the distances from the center of the earth .... Now if the lowest of these were very small and were so

---

[1] [Occasional lacunae in these propositions and corollaries indicate Newton's references to previous propositions in the *Principia*. Otherwise the text is complete.]

[m] [*Principia*, Bk. III. Proposition IV is the famous earth-moon test of the law of gravitation. As Cajori says, in his edition of the *Principia*, p. 663, this proposition is probably in its essentials the same computation that Newton made first in 1665-6. The Scholium to Proposition IV was added in the third edition of the *Principia* (1726).]

near the earth as almost to touch the tops of the highest mountains, the centripetal force thereof, retaining in its orbit, would be nearly equal to the weights of any terrestrial bodies that should be found upon the tops of those mountains, as may be known by the foregoing computation. Therefore, if the same little moon should be deserted by its centrifugal force that carries it through its orbit, and be disabled from going onward therein, it would descend to the earth; and that with the same velocity with which heavy bodies actually fall upon the tops of those very mountains, because of the equality of the forces that oblige them both to descend. And if the force by which the lowest moon would descend were different from gravity, and if the moon were to gravitate toward the earth, as we find terrestrial bodies do upon the tops of mountains, it would then descend with twice the velocity, as being impelled by both these forces conspiring together. Therefore, since both these forces, that is, the gravity of heavy bodies and the centripetal forces of the moons, are directed to the center of the earth and are similar and equal between themselves, they will (by Rules I and II) [n] have one and the same cause. And therefore the force which retains the moon in its orbit is that very force which we commonly call 'gravity,' because otherwise this little moon at the top of a mountain must either be without gravity or fall twice as swiftly as heavy bodies are wont to do.

SCHOLIUM [o]

The force which retains the celestial bodies in their orbits has been hitherto called 'centripetal force,' but it being now made plain that it can be no other than a gravitating force we shall hereafter call it 'gravity.' For the cause of that centripetal force which retains the moon in its orbit will extend itself to all the planets, by Rules I, II, and IV.

[n] [*I.e.*, the rules of reasoning in philosophy. See Part I.]
[o] [Scholium to Proposition V, *Principia*, Bk. III.]

PROPOSITION VI [p]

That all bodies gravitate toward every planet; and that the weights of bodies toward any one planet, at equal distances from the center of the planet, are proportional to the quantities of matter which they severally contain.

It has been now for a long time observed by others that all sorts of heavy bodies (allowance being made for the inequality of retardation which they suffer from a small power of resistance in the air) descend to the earth *from equal heights* in equal times, and that equality of times we may distinguish to a great accuracy by the help of pendulums. I tried experiments with gold, silver, lead, glass, sand, common salt, wood, water, and wheat. I provided two wooden boxes, round and equal; I filled the one with wood and suspended an equal weight of gold (as exactly as I could) in the center of oscillation of the other. The boxes, hanging by equal threads of 11 feet, made a couple of pendulums perfectly equal in weight and figure and equally receiving the resistance of the air. And, placing the one by the other, I observed them to play together forward and backward for a long time with equal vibrations. And therefore the quantity of matter in the gold . . . was to the quantity of matter in the wood as the action of the motive force (or *vis motrix*) upon all the gold to the action of the same upon all the wood, that is, as the weight of the one to the weight of the other; and the like happened in the other bodies. By these experiments, in bodies of the same weight, I could manifestly have discovered a difference of matter less than the thousandth part of the whole, had any such been. But, without all doubt, the nature of gravity toward the planets is the same as toward the earth. For, should we imagine our terrestrial bodies taken to the orbit of the moon and there, together with the moon, deprived of all motion, to be let go so as to fall together toward the earth, it is certain, from what we have demonstrated before, that in equal times they would describe equal spaces with the moon, and of consequence are to the moon, in quantity of matter, as their weights to its weight.

[p] [Proposition VI and Corollaries I, II, III, IV, and V, *Principia,* Bk. III.]

Moreover, since the satellites of Jupiter perform their revolutions in times which observe the 3/2th power of the proportion of their distances from Jupiter's center, their accelerative gravities toward Jupiter will be inversely as the squares of their distances from Jupiter's center, that is, equal at equal distances. And therefore these satellites, if supposed to fall *toward Jupiter* from equal heights, would describe equal spaces in equal times, in like manner as heavy bodies do on our earth. And, by the same argument, if the circumsolar planets were supposed to be let fall at equal distances from the sun, they would, in their descent toward the sun, describe equal spaces in equal times. But forces which equally accelerate unequal bodies must be as those bodies; that is to say, the weights of the planets *toward the sun* must be as their quantities of matter. Further, that the weights of Jupiter and of his satellites toward the sun are proportional to the several quantities of their matter appears from the exceedingly regular motions of the satellites. . . . For if some of those bodies were more strongly attracted to the sun in proportion to their quantity of matter than others, the motions of the satellites would be disturbed by that inequality of attraction. . . . If at equal distances from the sun any satellite, in proportion to the quantity of its matter, did gravitate toward the sun with a force greater than Jupiter in proportion to his, according to any given proportion, suppose of d to e, then the distance between the centers of the sun and of the satellite's orbit would be always greater than the distance between the centers of the sun and of Jupiter, nearly as the square root of that proportion, as by some computations I have found. And if the satellite did gravitate toward the sun with a force less in the proportion of e to d, the distance of the center of the satellite's orbit from the sun would be less than the distance of the center of Jupiter from the sun as the square root of the same proportion. Therefore if, at equal distances from the sun, the accelerative gravity of any satellite toward the sun were greater or less than the accelerative gravity of Jupiter toward the sun but by one 1/1,000 part of the whole gravity, the distance of the center of the satellite's orbit from the sun would be greater or less than the distance of Jupiter from the sun by one 1/2,000 part of the whole distance;

that is, by a fifth part of the distance of the utmost satellite from the center of Jupiter, an eccentricity of the orbit which would be very sensible. But the orbits of the satellites are concentric to Jupiter, and therefore the accelerative gravities of Jupiter and of all its satellites toward the sun are equal among themselves. And by the same argument the weights of Saturn and of his satellites toward the sun, at equal distances from the sun, are as their several quantities of matter; and the weights of the moon and of the earth toward the sun are either none or accurately proportional to the masses of matter which they contain. But some weight they have, by Corollaries I and III, Proposition V.

But further, the weights of all the parts of every planet toward any other planet are one to another as the matter in the several parts; for if some parts did gravitate more, others less, than for the quantity of their matter, then the whole planet, according to the sort of parts with which it most abounds, would gravitate more or less than in proportion to the quantity of matter in the whole. Nor is it of any moment whether these parts are external or internal; for if, for example, we should imagine the terrestrial bodies with us to be raised to the orbit of the moon, to be there compared with its body, if the weights of such bodies were to the weights of the external parts of the moon as the quantities of matter in the one and in the other respectively, but to the weights of the internal parts in a greater or less proportion, then likewise the weights of those bodies would be to the weight of the whole moon in a greater or less proportion, against what we have shown above.

COROLLARY I

Hence the weights of bodies do not depend upon their forms and textures; for if the weights could be altered with the forms, they would be greater or less, according to the variety of forms, in equal matter, altogether against experience.

COROLLARY II

Universally, all bodies about the earth gravitate toward the earth; and the weights of all, at equal distances from the earth's center, are as the quantities of matter which they severally contain. This is the quality of all bodies within the reach of our experiments, and therefore (by Rule III) to be affirmed of all bodies whatsoever. If the ether, or any other body, were either altogether void of gravity or were to gravitate less in proportion to its quantity of matter, then, because (according to Aristotle, Descartes, and others) there is no difference between that and other bodies but in mere form of matter, by a successive change from form to form, it might be changed at last into a body of the same condition with those which gravitate most in proportion to their quantity of matter; and, on the other hand, the heaviest bodies, acquiring the first form of that body, might by degrees quite lose their gravity. And therefore the weights would depend upon the forms of bodies and, with those forms, might be changed, contrary to what was proved in the preceding Corollary.

COROLLARY III

All spaces are not equally full; for if all spaces were equally full, then the specific gravity of the fluid which fills the region of the air, on account of the extreme density of the matter, would fall nothing short of the specific gravity of quicksilver or gold or any other the most dense body, and therefore neither gold nor any other body could descend in air, for bodies do not descend in fluids unless they are specifically heavier than the fluids. And if the quantity of matter in a given space can, by any rarefaction, be diminished, what should hinder a diminution to infinity?

COROLLARY IV

If all the solid particles of all bodies are of the same density and cannot be rarefied without pores, then a void, space, or vacuum must be granted. By 'bodies of the same density' I mean those whose inertias are in the proportion of their bulks.

COROLLARY V

The power of gravity is of a different nature from the power of magnetism, for the magnetic attraction is not as the matter attracted. Some bodies are attracted more by the magnet, others less; most bodies not at all. The power of magnetism in one and the same body may be increased and diminished, and is sometimes far stronger, for the quantity of matter, than the power of gravity; and in receding from the magnet decreases, not as the square, but almost as the cube of the distance, as nearly as I could judge from some rude observations.

ETHER AND GRAVITY [12]

From a letter to Robert Boyle [q]

Honored Sir,

I have so long deferred to send you my thoughts about the physical qualities we speak of that, did I not esteem myself obliged by promise, I think I should be ashamed to send them at all. The truth is, my notions about things of this kind are so indigested that I am not well satisfied myself in them; and what I am not satisfied in I can scarce esteem fit to be communicated to others, especially in natural philosophy, where there is no end of fancying. But because I am indebted to you, and yesterday met with a friend, Mr. Maulyverer, who told me he was going to London and intended to give you the trouble of a visit, I could not forbear to take the opportunity of conveying this to you by him.

It being only an explication of qualities which you desire of me, I shall set down my apprehensions in the form of suppositions as follows. And first, I suppose that there is diffused through all places an etherial substance, capable of contraction and dilatation,

q [*Opera Omnia* V, pp. 385-94. Also quoted in Brewster, *op. cit.*, Vol. I, pp. 409-19.]

strongly elastic, and, in a word, much like air in all respects, but far more subtle.

2. I suppose this ether pervades all gross bodies, but yet so as to stand rarer in their pores than in free spaces, and so much the rarer as their pores are less; and this I suppose (with others) to be the cause why light incident on those bodies is refracted toward the perpendicular, why two well-polished metals cohere in a receiver exhausted of air, why ☿ [mercury] stands sometimes up to the top of a glass pipe though much higher than thirty inches, and one of the main causes why the parts of all bodies cohere; also the cause of filtration and of the rising of water in small glass pipes above the surface of the stagnating water they are dipped into; for I suspect the ether may stand rarer, not only in the insensible pores of bodies, but even in the very sensible cavities of those pipes; and the same principle may cause menstruums to pervade with violence the pores of the bodies they dissolve, the surrounding ether, as well as the atmosphere, pressing them together.

3. I suppose the rarer ether within bodies and the denser without them not to be terminated in a mathematical superficies, but to grow gradually into one another; the external ether beginning to grow rarer and the internal to grow denser at some little distance from the superficies of the body, and running through all intermediate degrees of density in the intermediate spaces; and this may be the cause why light, in Grimaldo's experiment, passing by the edge of a knife or other opaque body is turned aside and as it were refracted, and by that refraction makes several colors....

4. When two bodies moving toward one another come near together, I suppose the ether between them to grow rarer than before and the spaces of its graduated rarity to extend further from the superficies of the bodies toward one another, and this by reason that the ether cannot move and play up and down so freely in the straight passage between the bodies as it could before they came so near together ....

5. Now, from the fourth supposition, it follows that when two bodies approaching one another come so near together as to make the ether between them begin to rarefy, they will begin to have a reluctance from being brought nearer together and an endeavor to

recede from one another, which reluctance and endeavor will increase as they come nearer together, because thereby they cause the interjacent ether to rarefy more and more. But at length, when they come so near together that the excess of pressure of the external ether which surrounds the bodies, above that of the rarefied ether which is between them, is so great as to overcome the reluctance which the bodies have from being brought together, then will that excess of pressure drive them with violence together and make them adhere strongly to one another, as was said in the second supposition. . . . But if the bodies come nearer together, so as to make the ether in the midway line . . . grow rarer than the surrounding ether, there will arise from the excess of density of the surrounding ether a compressure of the bodies toward one another which, when by the nearer approach of the bodies it becomes so great as to overcome the aforesaid endeavor the bodies have to recede from one another, they will then go toward one another and adhere together. And, on the contrary, if any power force them asunder to that distance where the endeavor to recede begins to overcome the endeavor to accede, they will again leap from one another. Now hence I conceive it is chiefly that a fly walks on water without wetting her feet, and consequently without touching the water; that two polished pieces of glass are not without pressure brought to contact, no, not though the one be plain, the other a little convex; that the particles of dust cannot by pressing be made to cohere, as they would do if they did but fully touch; that the particles of tinging substances and salts dissolved in water do not of their own accord concrete and fall to the bottom, but diffuse themselves all over the liquor and expand still more if you add more liquor to them. Also, that the particles of vapors, exhalations, and air do stand at a distance from one another and endeavor to recede as far from one another as the pressure of the incumbent atmosphere will let them; for I conceive the confused mass of vapors, air, and exhalations which we call the atmosphere to be nothing else but the particles of all sorts of bodies of which the earth consists, separated from one another and kept at a distance by the said principle. . . .

Nor does the size only, but the density of the particles also, conduce to the permanency of aerial substances; for the excess of density of the ether without such particles above that of the ether within them is still greater, which has made me sometimes think that the true permanent air may be of a metallic original, the particles of no substances being more dense than those of metals. This, I think, is also favored by experience, for I remember I once read in the *Philosophical Transactions* how M. Huygens at Paris found that the air made by dissolving salt of tartar would in two or three days' time condense and fall down again, but the air made by dissolving a metal continued without condensing or relenting in the least. If you consider, then, how by the continual fermentations made in the bowels of the earth there are aerial substances raised out of all kinds of bodies, all which together make the atmosphere, and that of all these the metallic are the most permanent, you will not perhaps think it absurd that the most permanent part of the atmosphere, which is the true air, should be constituted of these, especially since they are the heaviest of all other, and so must subside to the lower parts of the atmosphere and float upon the surface of the earth, and buoy up the lighter exhalations and vapors to float in greatest plenty above them. Thus, I say, it ought to be with the metallic exhalations raised in the bowels of the earth by the action of acid menstruums, and thus it is with the true permanent air; for this, as in reason it ought to be esteemed the most ponderous part of the atmosphere because the lowest, so it betrays its ponderosity by making vapors ascend readily in it, by sustaining mists and clouds of snow, and by buoying up gross and ponderous smoke. The air also is the most gross inactive part of the atmosphere, affording living things no nourishment, if deprived of the more tender exhalations and spirits that float in it; and what more inactive and remote from nourishment than metallic bodies?

I shall set down one conjecture more, which came into my mind now as I was writing this letter; it is about the cause of gravity. For this end I will suppose ether to consist of parts differing from one another in sublety by indefinite degrees; that in the pores of bodies there is less of the grosser ether, in proportion to the finer, than in open spaces; and consequently that in the great body of the earth there is much less of the grosser ether, in proportion to the finer, than in the regions of the air; and that yet the grosser ether in the air affects the upper regions of the earth, and the finer ether in the earth the lower regions of the air, in such a manner that from the top of the air to the surface of the earth, and again from the surface of the earth to the center thereof, the ether is insensibly finer and finer. Imagine now any body suspended in the air or lying on the earth, and the ether being by the hypothesis grosser in the pores which are in the upper parts of the body than in those which are in its lower parts, and that grosser ether being less apt to be lodged in those pores than the finer ether below, it will endeavor to get out and give way to the finer ether below, which cannot be without the bodies descending to make room above for it to go out into.

From this supposed gradual subtlety of the parts of ether some things above might be further illustrated and made more intelligible; but by what has been said you will easily discern whether in these conjectures there be any degree of probability, which is all I aim at. For my own part, I have so little fancy to things of this nature that, had not your encouragement moved me to it, I should never, I think, have thus far set pen to paper about them. What is amiss, therefore, I hope you will the more easily pardon in

Your most humble servant and honorer,

ISAAC NEWTON

Cambridge, February 28, 1678/9

6. COTES' PREFACE TO THE SECOND EDITION
OF THE *PRINCIPIA* [r]

We hereby present to the benevolent reader the long-awaited new edition of Newton's *Philosophy,* now greatly amended and increased. The principal contents of this celebrated work may be gathered from the [Table of Contents]. What has been added or modified is indicated in the author's Preface. There remains for us to add something relating to the method of this philosophy.

[r] [Cambridge, May 12, 1713. See Note 13, p. 198.]

Those who have treated of natural philosophy may be reduced to about three classes. Of these some have attributed to the several species of things specific and occult qualities, according to which the phenomena of particular bodies are supposed to proceed in some unknown manner. The sum of the doctrine of the schools derived from Aristotle and the Peripatetics is founded on this principle. They affirm that the several effects of bodies arise from the particular natures of those bodies. But whence it is that bodies derive those natures they don't tell us, and therefore they tell us nothing. And being entirely employed in giving names to things and not in searching into things themselves, they have invented, we may say, a philosophical way of speaking, but they have not made known to us true philosophy.

Others have endeavored to apply their labors to greater advantage by rejecting that useless medley of words. They assume that all matter is homogeneous, and that the variety of forms which is seen in bodies arises from some very plain and simple relations of the component particles. And by going on from simple things to those which are more compounded they certainly proceed right, if they attribute to those primary relations no other relations than those which Nature has given. But when they take a liberty of imagining at pleasure unknown figures and magnitudes, and uncertain situations and motions of the parts, and moreover of supposing occult fluids, freely pervading the pores of bodies, endued with an all-performing subtlety and agitated with occult motions, they run out into dreams and chimeras, and neglect the true constitution of things, which certainly is not to be derived from fallacious conjectures when we can scarce reach it by the most certain observations. Those who assume hypotheses as first principles of their speculations, although they afterward proceed with the greatest accuracy from those principles, may indeed form an ingenious romance, but a romance it will still be.

There is left then the third class, which possesses experimental philosophy. These indeed derive the causes of all things from the most simple principles possible; but then they assume nothing as a principle that is not proved by phenomena. They frame no hypotheses, nor receive them into philosophy otherwise than as questions whose truth may be disputed. They proceed therefore in a twofold method, synthetical and analytical. From some select phenomena they deduce by analysis the forces of Nature and the more simple laws of forces, and from thence by synthesis show the constitution of the rest. This is that incomparably best way of philosophizing which our renowned author most justly embraced in preference to the rest and thought alone worthy to be cultivated and adorned by his excellent labors. Of this he has given us a most illustrious example, by the explication of the System of the World, most happily deduced from the Theory of Gravity. That the attribute of gravity was found in all bodies others suspected or imagined before him, but he was the only and the first philosopher that could demonstrate it from appearances and make it a solid foundation to the most noble speculations.

I know indeed that some persons, and those of great name, too much prepossessed with certain prejudices, are unwilling to assent to this new principle and are ready to prefer uncertain notions to certain. It is not my intention to detract from the reputation of these eminent men; I shall only lay before the reader such considerations as will enable him to pass an equitable judgment in this dispute.

Therefore, that we may begin our reasoning from what is most simple and nearest to us, let us consider a little what is the nature of gravity in earthly bodies, that we may proceed the more safely when we come to consider it in the heavenly bodies that lie at the remotest distance from us. It is now agreed by all philosophers that all circumterrestrial bodies gravitate toward the earth. That no bodies having no weight are to be found is now confirmed by manifold experience. That which is relative levity is not true levity, but apparent only, and arises from the preponderating gravity of the contiguous bodies.

Moreover, as all bodies gravitate toward the earth, so does the earth gravitate again toward all bodies. That the action of gravity is mutual and equal on both sides is thus proved. Let the mass of the earth be divided into any two parts whatever, either equal or unequal; now if the weights of the parts toward each other were not mutually equal, the lesser weight would give way to the greater, and the two parts would move on together indefinitely in a right line toward that point to which the greater weight tends, which is altogether contrary to experience. Therefore we must say that the weights with which the parts tend to each other are equal; that is, that the action of gravity is mutual and equal in contrary directions.

The weights of bodies at equal distances from the center of the earth are as the quantities of matter in the bodies. This is inferred from the equal acceleration of all bodies that fall from a state of rest by their weights, for the forces by which unequal bodies are equally accelerated must be proportional to the quantities of the matter to be moved. Now that all falling bodies are equally accelerated appears from this that when the resistance of the air is taken away, as it is under an exhausted receiver of Mr. Boyle, they describe equal spaces in equal times; but this is yet more accurately proved by the experiments with pendulums.

The attractive forces of bodies at equal distances are as the quantities of matter in the bodies. For since bodies gravitate toward the earth and the earth again toward bodies with equal moments, the weight of the earth toward each body or the force with which the body attracts the earth will be equal to the weight of the same body toward the earth. But this weight was shown to be as the quantity of matter in the body, and therefore the force with which each body attracts the earth, or the absolute force of the body, will be as the same quantity of matter.

Therefore the attractive force of the entire bodies arises from and is composed of the attractive forces of the parts, because, as was just shown, if the bulk of the matter be augmented or diminished, its power is proportionately augmented or diminished. We must therefore conclude that the action of the earth is composed of the united actions of its parts, and therefore that all terrestrial bodies must attract one another mutually, with absolute forces that are as the matter attracting. This is the nature of gravity upon earth; let us now see what it is in the heavens.

That every body continues in its state either of rest or of moving uniformly in a right line, unless so far as it is compelled to change that state by external force, is a law of Nature universally received by all philosophers. But it follows from this that bodies which move in curved lines, and are therefore continually bent from the right lines that are tangents to their orbits, are retained in their curvilinear paths by some force continually acting. Since, then, the planets move in curvilinear orbits, there must be some force operating by the incessant actions of which they are continually made to deflect from the tangents.

Now it is evident from mathematical reasoning, and rigorously demonstrated, that all bodies that move in any curved line described in a plane and which, by a radius drawn to any point, whether at rest or moved in any manner, describe areas about that point proportional to the times are urged by forces directed toward that point. This must therefore be granted. Since, then, all astronomers agree that the primary planets describe about the sun and the secondary about the primary areas proportional to the times, it follows that the forces by which they are continually turned aside from the rectilinear tangents and made to revolve in curvilinear orbits are directed toward the bodies that are placed in the centers of the orbits. This force may therefore not improperly be called centripetal in respect of the revolving body, and in respect of the central body attractive, from whatever cause it may be imagined to arise.

Moreover, it must be granted, as being mathematically demonstrated, that if several bodies revolve with an equable motion in concentric circles and the squares of the periodic times are as the cubes of the distances from the common center, the centripetal forces will be inversely as the squares of the distances. Or, if bodies revolve in orbits that are very nearly circular and the apsides of the orbits are at rest, the centripetal forces of the revolving bodies will be inversely as the squares of the distances. That both these facts hold for all the planets, all astronomers agree. Therefore the centripetal forces of all the planets are inversely as the squares of the distances from the centers of their orbits. If any should object that the apsides of the planets, and especially of the moon, are not perfectly at rest, but are carried progressively with a slow kind of motion, one may give this answer, that, though we should grant that this very slow motion arises from a slight deviation of the centripetal force from the law of the square of the distance, yet we are able to compute mathematically the quantity of that aberration and find it perfectly insensible. For even the ratio of the lunar centripetal force itself, which is the most irregular of them all, will vary inversely as a power a little greater than the square of the distance, but will be well-nigh sixty times nearer to the square than to the cube of the distance. But we may give a truer answer by saying that this progression of the apsides arises, not from a deviation from the law of inverse squares of the distance, but from a quite different cause, as is most admirably shown in this work. It is certain then that the centripetal forces with which the primary planets tend to the sun and the secondary planets to their primary are accurately as the inverse squares of the distances.

From what has been hitherto said, it is plain that the planets are retained in their orbits by some force continually acting upon them; it is plain that this force is always directed toward the centers of their orbits; it is plain that its intensity is increased in its approach and is decreased in its recession from the center, and that it is increased in the same ratio in which the square of the distance is diminished and decreased in the same ratio in which the square of the distance is augmented. Let us now see whether, by making a comparison between the centripetal forces of the planets and the force of gravity, we may not by chance find them to be of the same kind. Now, they will be of the same kind if we find on both sides the same laws and the same attributes. Let us then first consider the centripetal force of the moon, which is nearest to us.

The rectilinear spaces which bodies let fall from rest describe in a given time at the very beginning of the motion, when the bodies are urged by any forces whatsoever, are proportional to the forces. This appears from mathematical reasoning. Therefore the centripetal force of the moon revolving in its orbit is to the force of gravity at the surface of the earth as the space which in a very small interval of time the moon, deprived of all its circular force and descending by its centripetal force toward the earth, would describe is to the space which a heavy body would describe when falling by the force of its gravity near to the earth in the same small interval of time. The first of these spaces is equal to the versed sine of the arc described by the moon in the same time, because that versed sine measures the translation of the moon from the tangent produced by the centripetal force, and therefore may be computed if the periodic time of the moon and its distance from the center of the earth are given. The last space is found by experiments with pendulums, as Mr. Huygens has shown. Therefore, by making a calculation, we shall find that the first space is to the latter, or the centripetal force of the moon revolving in its orbit will be to the force of gravity at the surface of the earth, as the square of the semidiameter of the earth to the square of the semidiameter of the orbit. But, by what was shown before, the very same ratio holds between the centripetal force of the moon revolving in its orbit and the centripetal force of the moon near the surface of the earth. Therefore the centripetal force near the surface of the earth is equal to the force of gravity. Therefore these are not two different forces, but one and the same; for if they were different, these forces united would cause bodies to descend to the earth with twice the velocity they would fall with by the force of gravity alone. Therefore it is plain that the centripetal force, by which the moon is continually either impelled or attracted out of the tangent and retained in its orbit, is the very force of terrestrial gravity reaching up to the moon. And it is very reasonable to believe that this force should extend itself to vast distances, since upon the tops of the highest mountains we find no sensible diminution of it. Therefore the moon gravitates toward the earth; but, on the other hand, the earth by a mutual action equally gravitates toward the moon, which is also abundantly confirmed in this philosophy, where the tides in the sea and the precession of the equinoxes are treated of, which arise from the action both of the moon and of the sun upon the earth. Hence, lastly, we discover by what law the force of gravity decreases at great distances from the earth. For since gravity is noways different from the moon's centripetal force, and this is inversely proportional to the square of the distance, it follows that it is in that very ratio that the force of gravity decreases.

Let us now go on to the other planets. Because the revolutions of the primary planets about the sun and of the secondary about Jupiter and Saturn are phenomena of the same kind with the revolution of the moon about the earth, and because it has been moreover demonstrated that the centripetal forces of the primary planets are directed toward the center of the sun and those of the secondary toward the centers of Jupiter and Saturn in the same manner as the centripetal force of the moon is directed toward the center of the earth, and since, besides, all these forces are inversely as the squares of the distances from the centers, in the same manner as the centripetal force of the moon is as the square of the distance from the earth, we must of course conclude that the nature of all is the same. Therefore as the moon gravitates toward the earth and the earth again toward the moon, so also all the secondary planets will gravitate toward their primary, and the primary planets again toward their secondary, and so all the primary toward the sun, and the sun again toward the primary.

Therefore the sun gravitates toward all the planets, and all the planets toward the sun. For the secondary planets, while they accompany the primary, revolve the meanwhile with the primary about the sun. Therefore, by the same argument, the planets of both kinds gravitate toward the sun and the sun toward them. That the secondary planets gravitate toward the sun is moreover abundantly clear from the inequalities of the moon, a most accurate theory of which, laid open with a most admirable sagacity, we find explained in the Third Book of this work.

That the attractive force of the sun is propagated on all sides to prodigious distances and is diffused to every part of the wide space that surrounds it is most evidently shown by the motion of the comets, which, coming from places immensely distant from the sun, approach very near to it, and sometimes so near that in their perihelia they almost touch its body. The theory of these bodies was altogether unknown to astronomers till, in our own times, our excellent author most happily discovered it and demonstrated the truth of it by most certain observations. So that it is now apparent that the comets move in conic sections having their foci in the sun's center, and by radii drawn to the sun describe areas proportional to the times. But from these phenomena it is manifest and mathematically demonstrated that those forces by which the comets are retained in their orbits are directed toward the sun and are inversely proportional to the squares of the distances from its center. Therefore the comets gravitate toward the sun, and therefore the attractive force of the sun not only acts on the bodies of the planets, placed at given distances and very nearly in the same plane, but reaches also the comets in the most different parts of the heavens and at the most different distances. This, therefore, is the nature of gravitating bodies, to exert their force at all distances to all other gravitating bodies. But from thence it follows that all the planets and comets attract one another mutually and gravitate toward one another, which is also confirmed by the perturbation of Jupiter and Saturn, observed by astronomers and arising from the mutual actions of these two planets upon each other, as also from that very slow motion of the apsides, above taken notice of, which arises from a like cause.

We have now proceeded so far that it must be acknowledged that the sun and the earth, and all the heavenly bodies attending the sun, attract one another mutually. Therefore all the least particles of matter in every one must have their several attractive forces proportional to their quantities of matter, as was shown above of the terrestrial bodies. At different distances these forces will be also inversely as the squares of their distances, for it is mathematically demonstrated that globes attracting according to this law are composed of particles attracting according to the same law.

The foregoing conclusions are grounded on this axiom which is received by all philosophers, namely, that effects of the same kind, whose known properties are the same, take their rise from the same causes and have the same unknown properties also. For if gravity be the cause of the descent of a stone in Europe, who doubts that it is also the cause of the same descent in America? If there is a mutual gravitation between a stone and the earth in Europe, who will deny the same to be mutual in America? If in Europe the attractive force of a stone and the earth is composed of the attractive forces of the parts, who will deny the like composition in America? If in Europe the attraction of the earth be propagated to all kinds of bodies and to all distances, why may we not say that it is propagated in like manner in America? All philosophy is founded on this rule; for if that be taken away, we can affirm nothing as a general truth. The constitution of particular things is known by observations and experiments; and when that is done, no general conclusion of the nature of things can thence be drawn except by this rule.

Since, then, all bodies, whether upon earth or in the heavens, are heavy, so far as we can make any experiments or observations concerning them, we must certainly allow that gravity is found in all bodies universally. And in like manner as we ought not to suppose that any bodies can be otherwise than extended, movable, or impenetrable, so we ought not to conceive that any bodies can be otherwise than heavy. The extension, mobility, and impenetrability of bodies become known to us only by experiments, and in the very same manner their gravity becomes known to us. All bodies upon which we can make any observations are extended, movable, and impenetrable; and thence we conclude all bodies, and those concerning which we have no observations, are extended and movable and impenetrable. So all bodies on which we can make observations we find to be heavy; and thence we conclude all bodies, and those we have no observations of, to be heavy also. If anyone should say that the bodies of the fixed stars are not heavy because their gravity is not yet observed, they may say for the same reason that they are neither extended nor movable nor impenetrable because these properties of the fixed stars are not yet observed. In short, either gravity must have a place among the primary qualities of all bodies, or extension, mobility, and impenetrability must not. And if the nature of things is not rightly explained by the gravity of bodies, it will not be rightly explained by their extension, mobility, and impenetrability.

Some I know disapprove this conclusion and mutter something about occult qualities. They continually are cavilling with us that gravity is an occult property, and occult causes are to be quite banished from philosophy. But to this the answer is easy: that those are indeed occult causes whose existence is occult, and imagined but not proved, but not those whose real existence is clearly demonstrated by observations. Therefore gravity can by no means be called an occult cause of the celestial motions, because it is plain from the phenomena that such a power does really exist. Those rather have recourse to occult causes who set imaginary vortices of a matter entirely fictitious and imperceptible by our senses to direct those motions.

But shall gravity be therefore called an occult cause and thrown out of philosophy because the cause of gravity is occult and not yet discovered? Those who affirm this should be careful not to fall into an absurdity that may overturn the foundations of all philosophy. For causes usually proceed in a continued chain from those that are more compounded to those that are more simple; when we are arrived at the most simple cause, we can go no farther. Therefore no mechanical account or explanation of the most simple cause is to be expected or given; for if it could be given, the cause were not the most simple. These most simple causes will you then call occult and reject them? Then you must reject those that immediately depend upon them and those which depend upon these last, till philosophy is quite cleared and disencumbered of all causes.

Some there are who say that gravity is preternatural and call it a perpetual miracle. Therefore they would have it rejected, because preternatural causes have no place in physics. It is hardly worth while to spend time in answering this ridiculous objection which overturns all philosophy. For either they will deny gravity to be in bodies, which cannot be said, or else they will therefore call it preternatural because it is not produced by the other properties of bodies, and therefore not by mechanical causes. But certainly there are primary properties of bodies; and these, because they are primary, have no dependence on the others. Let them consider whether all these are not in like manner preternatural and in like manner to be rejected, and then what kind of philosophy we are like to have.

Some there are who dislike this celestial physics because it contradicts the opinions of Descartes and seems hardly to be reconciled with them. Let these enjoy their own opinion, but let them act fairly and not deny the same liberty to us which they demand for themselves. Since the Newtonian philosophy appears true to us, let us have the liberty to embrace and retain it, and to follow causes proved by phenomena, rather than causes only imagined and not yet proved. The business of true philosophy is to derive the natures of things from causes truly existent and to inquire after those laws on which the Great Creator actually chose to found this most beautiful Frame of the World, not those by which he might have done the same had he so pleased. It is reasonable enough to suppose that from several causes, somewhat differing from one another, the same effect may arise, but the true cause will be that from which it truly and actually does arise; the others have no place in true philosophy. The same motion of the hour hand in a clock may be occasioned either by a weight hung or a spring shut up within. But if a certain clock should be really moved with a weight, we should laugh at a man that would suppose it moved by a spring, and from that principle, suddenly taken up without further examination, should go about to explain the motion of the index; for certainly the way he ought to have taken would have been actually to look into the inward parts of the machine, that he might find the true principle of the proposed motion. The like judgment ought to be made of those philosophers who will have the heavens to be filled with a most subtle matter which is continually carried round in vortices. For if they could explain the phenomena ever so accurately by their hypotheses, we could not yet say that they have discovered true philosophy and the true causes of the celestial motions, unless they could either demonstrate that those causes do actually exist, or at least that no others do exist. Therefore, if it be made clear that the attraction of all bodies is a property actually existing *in rerum natura,* and if it be also shown how the motions of the celestial bodies may be solved by that property, it would be very impertinent for anyone to object that these motions ought to be accounted for by vortices, even though we should allow such an explication of those motions to be possible. But we allow no such thing; for the phenomena can by no means be accounted for by vortices, as our author has abundantly proved from the clearest reasons. So that men must be strangely fond of chimeras who can spend their time so idly as in patching up a ridiculous figment and setting it off with new comments of their own.

If the bodies of the planets and comets are carried round the sun in vortices, the bodies so carried, and the parts of the vortices next surrounding them, must be carried with the same velocity and the same direction, and have the same density and the same inertia, answering to the bulk of the matter. But it is certain, the planets and comets, when in the very same parts of the heavens, are carried with various velocities and various directions. Therefore it necessarily follows that those parts of the celestial fluid which are at the same distances from the sun, must revolve at the same time with different velocities in different directions; for one kind of velocity and direction is required for the motion of the planets, and another for that of the comets. But since this cannot be accounted for, we must either say that all celestial bodies are not carried about by vortices, or else that their motions are derived, not from one and the same vortex, but from several distinct ones which fill and pervade the spaces round about the sun.

But if several vortices are contained in the same space and are supposed to penetrate one another, and to revolve with different motions, then because these motions must agree with those of the bodies carried about by them, which are perfectly regular, and performed in conic sections which are sometimes very eccentric, and sometimes nearly circles, one may very reasonably ask how it comes to pass that these vortices remain entire and have suffered no manner of perturbation in so many ages from the actions of the conflicting matter. Certainly if these fictitious motions are more compounded and harder to be accounted for than the true motions of the planets and comets, it seems to no purpose to admit them into philosophy, since every cause ought to be more simple than its effect. Allowing men to indulge their own fancies, suppose any man should affirm that the planets and comets are surrounded with atmospheres like our earth, which hypothesis seems more reasonable than that of vortices; let him then affirm that these atmospheres, by their own nature, move about the sun and describe conic sections, which motion is much more easily conceived than that of the vortices penetrating one another; lastly, that the planets and comets are carried about the sun by these atmospheres of theirs: and then applaud his own sagacity in discovering the causes of the celestial motions. He that rejects this fable must also reject the other, for two drops of water are not more like than this hypothesis of atmospheres and that of vortices.

Galileo has shown that, when a stone projected moves in a parabola, its deflection into that curve from its rectilinear path is occasioned by the gravity of the stone toward the earth, that is, by an occult quality. But now somebody, more cunning than he, may come to explain the cause after this manner. He will suppose a certain subtle matter, not discernible by our sight, our touch, or any other of our senses, which fills the spaces which are near and contiguous to the surface of the earth, and that this matter is carried with different directions, and various and often contrary motions, describing parabolic curves. Then see how easily he may account for the deflection of the stone above spoken of. The stone, says he, floats in this subtle fluid, and following its motion cannot choose but describe the same figure. But the fluid moves in parabolic curves, and therefore the stone must move in a parabola, of course. Would not the acuteness of this philosopher be thought very extraordinary who could deduce the appearances of Nature from mechanical causes, matter and motion, so clearly that the meanest man may understand it? Or indeed should not we smile to see this new Galileo taking so much mathematical pains to introduce occult qualities into philosophy, from whence they have been so happily excluded? But I am ashamed to dwell so long upon trifles.

The sum of the matter is this: the number of the comets is certainly very great; their motions are perfectly regular and observe the same laws with those of the planets. The orbits in which they move are conic sections, and those very eccentric. They move every way toward all parts of the heavens and pass through the planetary regions with all possible freedom, and their motion is often contrary to the order of the signs. These phenomena are most evidently confirmed by astronomical observations and cannot be accounted for by vortices. Nay, indeed, they are utterly irreconcilable with the vortices of the planets. There can be no room for the motions of the comets, unless the celestial spaces be entirely cleared of that fictitious matter.

For if the planets are carried about the sun in vortices, the parts of the vortices which immediately surround every planet must be of the same density with the planet, as was shown above. Therefore all the matter contiguous to the perimeter of the earth's orbit must be of the same density as the earth. But this great orb and the orb of Saturn must have either an equal or a greater density. For to make the constitution of the vortex permanent, the parts of less density must lie near the center and those of greater density must go farther from it. For since the periodic times of the planets vary as the $3/2$th powers of their distances from the sun, the periods of the parts of the vortices must also preserve the same ratio. Thence it will follow that the centrifugal forces of the parts of the vortex must be inversely as the squares of their distances. Those parts, therefore, which are more remote from the center endeavor to recede from it with less force; whence, if their density be deficient, they must yield to the greater force with which the parts that lie nearer the center endeavor to ascend. Therefore the denser parts will ascend and those of less density will descend, and there will be a mutual change of places, till all the fluid matter in the whole vortex be so adjusted and disposed that, being reduced to an equilibrium, its parts become quiescent. If two fluids of different density be contained in the same vessel, it will certainly come to pass that the fluid of greater density will sink the lower; and by a like reasoning it follows that the denser parts of the vortex, by their greater centrifugal force, will ascend to the higher places. Therefore all that far greater part of the vortex which lies without the earth's orb will have a density, and by consequence an inertia, answering to the bulk of the matter, which cannot be less than the density and inertia of the earth. But from hence will arise a mighty resistance to the passage of the comets, such as must be very sensible, not to say enough to put a stop to and absorb their motions entirely. But it appears from the perfectly regular motion of the comets that they suffer no resistance that is in the least sensible, and therefore that they do not meet with matter of any kind that has any resisting force or, by consequence, any density or inertia. For the resistance of mediums arises either from the inertia of the matter of the fluid or from its want of lubricity. That which arises from the want of lubricity is very small and is scarcely observable in the fluids commonly known, unless they be very tenacious like oil and honey. The resistance we find in air, water, quicksilver, and the like fluids that are not tenacious is almost all of the first kind and cannot be diminished by a greater degree of subtlety, if the density and inertia to which this resistance is proportional remains, as is most evidently demonstrated by our author in his noble theory of resistances in Book II.

Bodies in going on through a fluid communicate their motion to the ambient fluid by little and little, and by that communication lose their own motion and, by losing it, are retarded. Therefore the retardation is proportional to the motion communicated, and the communicated motion, when the velocity of the moving body is given, is as the density of the fluid, and therefore the retardation or resistance will be as the same density of the fluid; nor can it be taken away, unless the fluid, coming about to the hinder parts of the body, restore the motion lost. Now this cannot be done unless the impression of the fluid on the hinderparts of the body be equal to the impression of the foreparts of the body on the fluid, that is, unless the relative velocity with which the fluid pushes the body behind is equal to the velocity with which the body pushes the fluid; that is, unless the absolute velocity of the recurring fluid be twice as great as the absolute velocity with which the fluid is driven forward by the body, which is impossible. Therefore the resistance of fluids arising from their inertia can by no means be taken away. So that we must conclude that the celestial fluid has no inertia because it has no resisting force; that it has no force to communicate motion with because it has no inertia; that it has no force to produce any change in one or more bodies because it has no force wherewith to communicate motion; that it has no manner of efficacy because it has no faculty wherewith to produce any change of any kind. Therefore certainly this hypothesis may be justly called ridiculous and unworthy of a philosopher, since it is altogether without foundation and does not in the least serve to explain the nature of things. Those who would have the heavens filled with a fluid matter, but suppose it void of any inertia, do indeed in words deny a vacuum but allow it in fact. For since a fluid matter of that kind can noways be distinguished from empty space, the dispute is now about the names and not the natures of things. If any are so fond of matter that they will by no means admit of a space void of body, let us consider where they must come at last.

For either they will say that this constitution of a world everywhere full was made so by the will of God to this end that the operations of Nature might be assisted everywhere by a subtle ether pervading and filling all things, which cannot be said, however, since we have shown from the phenomena of the comets that this ether is of no efficacy at all; or they will say that it became so by the same will of God for some unknown end, which ought not be said because, for the same reason, a different constitution may be as well supposed; or, lastly, they will not say that it was caused by the will of God, but by some necessity of its nature. Therefore they will at last sink into the mire of that infamous herd who dream that all things are governed by fate and not by providence, and that matter exists by the necessity of its nature always and everywhere, being infinite and eternal. But supposing these things, it must be also everywhere uniform; for variety of forms is entirely inconsistent with necessity. It must be also unmoved; for if it be necessarily moved in any determinate direction, with any determinate velocity, it will by a like necessity be moved in a different direction with a different velocity: but it can never move in different directions with different velocities; therefore it must be unmoved. Without all doubt this world, so diversified with that variety of forms and motions we find in it, could arise from nothing but the perfectly free will of God directing and presiding over all.

From this fountain it is that those laws which we call the laws of Nature have flowed, in which there appear many traces indeed of the most wise contrivance, but not the least shadow of necessity. These, therefore, we must not seek from uncertain conjectures, but learn them from observations and experiments. He who is presumptuous enough to think that he can find the true principles of physics and the laws of natural things by the force alone of his own mind and the internal light of his reason must either suppose that the world exists by necessity and by the same necessity follows the laws proposed or, if the order of Nature was established by the will of God, that himself, a miserable reptile, can tell what was fittest to be done. All sound and true philosophy is founded on the appearances of things; and if these phenomena inevitably draw us, against our wills, to such principles as most clearly manifest to us the most excellent counsel and supreme dominion of the All-wise and Almighty Being, they are not therefore to be laid aside because some men may perhaps dislike them. These men may call them miracles or occult qualities, but names maliciously given ought not to be a disadvantage to the things themselves, unless these men will say at last that all philosophy ought to be founded in atheism. Philosophy must not be corrupted in compliance with these men, for the order of things will not be changed.

Fair and equal judges will therefore give sentence in favor of this most excellent method of philosophy, which is founded on experiments and observations. And it can hardly be said or imagined what light, what splendor, has accrued to that method from this admirable work of our illustrious author, whose happy and sublime genius, resolving the most difficult problems and reaching to discoveries of which the mind of man was thought incapable before, is deservedly admired by all those who are somewhat more than superficially versed in these matters. The gates are now set open, and by the passage he has revealed we may freely enter into the knowledge of the hidden secrets and wonders of natural things. He has so clearly laid open and set before our eyes the most beautiful frame of the System of the World that if King Alphonso were now alive he would not complain for want of the graces either of simplicity or of harmony in it. Therefore we may now more nearly behold the beauties of Nature and entertain ourselves with the delightful contemplation, and, which is the best and most valuable fruit of philosophy, be thence incited the more profoundly to reverence and adore the great Maker and Lord of all.

He must be blind who, from the most wise and excellent contrivances of things, cannot see the infinite wisdom and goodness of their Almighty Creator, and he must be mad and senseless who refuses to acknowledge them.

Newton's distinguished work will be the safest protection against the attacks of atheists, and nowhere more surely than from this quiver can one draw forth missiles against the band of godless men. This was felt long ago and first surprisingly demonstrated in learned English and Latin discourses by Richard Bentley, who, excelling in learning and distinguished as a patron of the highest arts, is a great ornament of his century and of our academy, the most worthy and upright Master of our Trinity College. To him in many ways I must express my indebtedness. And you too, benevolent reader, will not withhold the esteem due him. For many years an intimate friend of the celebrated author (since he aimed not only that the author should be esteemed by those who come after, but also that these uncommon writings should enjoy distinction among the literati of the world), he cared both for the reputation of his friend and for the advancement of the sciences. Since copies of the previous edition were very scarce and held at high prices, he persuaded, by frequent entreaties and almost by chidings, the splendid man, distinguished alike for modesty and for erudition, to grant him permission for the appearance of his new edition, perfected throughout and enriched by new parts, at his expense and under his supervision. He assigned to me, as he had a right, the not unwelcome task of looking after the corrections as best I could.

ROGER COTES
Fellow of Trinity College
Plumian Professor of Astronomy
and Experimental Philosophy.

Cambridge, May 12, 1713

V. Questions from the *Optics*[a]

. . . I shall conclude with proposing only some Queries, in order to a further search to be made by others.

Query 1. Do not bodies act upon light at a distance, and by their action bend its rays; and is not this action (*ceteris paribus*) strongest at the least distance?

Qu. 2. Do not the rays which differ in refrangibility differ also in flexibility; and are they not by their different inflections separated from one another, so as after separation to make the colors in the three fringes above described? And after what manner are they inflected to make those fringes?

Qu. 3. Are not the rays of light, in passing by the edges and sides of bodies, bent several times backward and forward, with a motion like that of an eel? And do not the three fringes of colored light above mentioned arise from three such bendings?

Qu. 4. Do not the rays of light which fall upon bodies and are reflected or refracted begin to bend before they arrive at the bodies; and are they not reflected, refracted, and inflected by one and the same principle, acting variously in various circumstances?

Qu. 5. Do not bodies and light act mutually upon one another; that is to say, bodies upon light in emitting, reflecting, refracting and inflecting it, and light upon bodies for heating them and putting their parts into a vibrating motion wherein heat consists?

Qu. 6. Do not black bodies conceive heat more easily from light than those of other colors do, by reason that the light falling on them is not reflected outward but enters the bodies, and is often reflected and refracted within them, until it be stifled and lost?

Qu. 7. Is not the strength and vigor of the action between light and sulphureous bodies observed above one reason why sul-

a [*Opticks: or, a Treatise of the Reflections, Refractions, Inflections, and Colours of Light* (4th ed., corrected, London, 1730), Bk. III, pp. 313-82. See Note 14, p. 201.]

135

phureous bodies take fire more readily and burn more vehemently than other bodies do?

Qu. 8. Do not all fixed bodies, when heated beyond a certain degree, emit light and shine; and is not this emission performed by the vibrating motions of their parts? And do not all bodies which abound with terrestrial parts, and especially with sulphureous ones, emit light as often as those parts are sufficiently agitated, whether that agitation be made by heat or by friction or percussion or putrefaction, or by any vital motion or any other cause? As for instance: sea water in a raging storm; quicksilver agitated *in vacuo;* the back of a cat or neck of a horse obliquely struck or rubbed in a dark place; wood, flesh, and fish while they putrefy; vapors arising from putrefied waters, usually called *ignes fatui;* stacks of moist hay or corn growing hot by fermentation; glowworms and the eyes of some animals by vital motions; the vulgar phosphorus agitated by the attrition of any body or by the acid particles of the air; amber and some diamonds by striking, pressing, or rubbing them; scrapings of steel struck off with a flint; iron hammered very nimbly till it become so hot as to kindle sulphur thrown upon it; the axletrees of chariots taking fire by the rapid rotation of the wheels; and some liquors mixed with one another whose particles come together with an impetus, as oil of vitriol distilled from its weight of niter and then mixed with twice its weight of oil of anniseeds. So also a globe of glass about 8 or 10 inches in diameter, being put into a frame where it may be swiftly turned around its axis, will in turning shine where it rubs against the palm of one's hand applied to it; and if at the same time a piece of white paper or white cloth, or the end of one's finger, be held at the distance of about a quarter of an inch or half an inch from that part of the glass where it is most in motion, the electric vapor which is excited by the friction of the glass against the hand will, by dashing against the white paper, cloth, or finger, be put into such an agitation as to emit light and make the white paper, cloth, or finger appear lucid like a glowworm, and in rushing out of the glass will sometimes push against the finger so as to be felt. And the same things have been found by rubbing a long and large cylinder of glass or amber with a paper held in one's hand and continuing the friction till the glass grew warm.

Qu. 9. Is not fire a body heated so hot as to emit light copiously? For what else is a red-hot iron than fire? And what else is a burning coal than red-hot wood?

Qu. 10. Is not flame a vapor, fume, or exhalation heated red-hot, that is, so hot as to shine? For bodies do not flame without emitting a copious fume, and this fume burns in the flame. The *ignis fatuus* is a vapor shining without heat, and is there not the same difference between this vapor and flame as between rotten wood shining without heat and burning coals of fire? In distilling hot spirits, if the head of the still be taken off, the vapor which ascends out of the still will take fire at the flame of a candle and turn into flame, and the flame will run along the vapor from the candle to the still. Some bodies heated by motion or fermentation, if the heat grow intense, fume copiously, and if the heat be great enough the fumes will shine and become flame. Metals in fusion do not flame for want of a copious fume, except spelter, which fumes copiously and thereby flames. All flaming bodies, as oil, tallow, wax, wood, fossil coals, pitch, sulphur, by flaming, waste and vanish into burning smoke, which smoke, if the flame be put out, is very thick and visible and sometimes smells strongly, but in the flame loses its smell by burning; and according to the nature of the smoke the flame is of several colors, as that of sulphur blue, that of copper opened with sublimate green, that of tallow yellow, that of camphire white. Smoke passing through flame cannot but grow red-hot, and red-hot smoke can have no other appearance than that of flame. When gunpowder takes fire, it goes away into flaming smoke. For the charcoal and sulphur easily take fire and set fire to the niter, and the spirit of the niter being thereby rarified into vapor rushes out with explosion, much after the manner that the vapor of water rushes out of an aeolipile; the sulphur also being volatile is converted into vapor and augments the explosion. And the acid vapor of the sulphur (namely that which distils under a bell into oil of sulphur) entering violently into the fixed body of the niter sets loose the spirit of the niter and excites a great fermentation whereby the heat is further augmented and the fixed body of the niter is also rarified into fume, and the explosion is thereby made more vehement and quick. For if salt of tartar be mixed with gunpowder, and that mixture be warmed till it takes fire, the explosion will be more violent and quick than that of gunpowder alone, which cannot proceed from any other cause than the action of the vapor of the gunpowder upon the salt of tartar whereby that salt is rarified. The explosion of gunpowder arises therefore from the violent action whereby all the mixture being quickly and vehemently heated is rarified and converted into fume and vapor, which vapor, by the violence of that action becoming so hot as to shine, appears in the form of flame.

Qu. 11. Do not great bodies conserve their heat the longest, their parts heating one another; and may not great dense and fixed bodies, when heated beyond a certain degree, emit light so copiously as by the emission and reaction of its light and the reflections and refractions of its rays within its pores, to grow still hotter, till it comes to a certain period of heat such as is that of the sun? And are not the sun and fixed stars great earths vehemently hot, whose heat is conserved by the greatness of the bodies and the mutual action and reaction between them, and the light which they emit; and whose parts are kept from fuming away, not only by their fixity, but also by the vast weight and density of the atmospheres incumbent upon them and very strongly compressing them, and condensing the vapors and exhalations which arise from them? For if water be made warm in any pellucid vessel emptied of air, that water in the vacuum will bubble and boil as vehemently as it would in the open air in a vessel set upon the fire till it conceives a much greater heat. For the weight of the incumbent atmosphere keeps down the vapors and hinders the water from boiling until it grows much hotter than is requisite to make it boil *in vacuo*. Also a mixture of tin and lead being put upon a red-hot iron *in vacuo* emits a fume and flame, but the same mixture in the open air, by reason of the incumbent atmosphere, does not so much as emit any fume which can be perceived by sight. In like manner the great weight of the atmosphere which lies upon the globe of the sun may hinder bodies there from rising up and going away from the sun in the form of vapors and fumes, unless by means of a far greater heat than that which on the surface of our earth would very easily turn them into vapors and fumes. And the same great weight may condense those vapors and exhalations as soon as they shall at any time begin to ascend from the sun and make them presently fall back again into him, and by that action increase his heat much after the manner that in our earth the air increases the heat of a culinary fire. And the same weight may hinder the globe of the sun from being diminished, unless by the emission of light and a very small quantity of vapors and exhalations.

Qu. 12. Do not the rays of light, in falling upon the bottom of the eye, excite vibrations in the *tunica retina*? Which vibrations, being propagated along the solid fibers of the optic nerves into the brain, cause the sense of seeing. For because dense bodies conserve their heat a long time, and the densest bodies conserve their heat the longest, the vibrations of their parts are of a lasting nature and therefore may be propagated along solid fibers of uniform dense matter to a great distance, for conveying into the brain the impressions made upon all the organs of sense. For that motion which can continue long in one and the same part of a body can be propagated a long way from one part to another, supposing the body homogeneal, so that the motion may not be reflected, refracted, interrupted, or disordered by any unevenness of the body.

Qu. 13. Do not several sorts of rays make vibrations of several bignesses which, according to their bignesses, excite sensations of several colors, much after the manner that the vibrations of the air, according to their several bignesses, excite sensations of several sounds? And particularly do not the most refrangible rays excite the shortest vibrations for making a sensation of deep violet, the least refrangible the largest for making a sensation of deep red; and the several intermediate sorts of rays vibrations of several intermediate bignesses to make sensations of the several intermediate colors?

Qu. 14. May not the harmony and discord of colors arise from the proportions of the vibrations propagated through the fibers of the optic nerves into the brain, as the harmony and discord of sounds arise from the proportions of the vibrations of the air? For some colors, if they be viewed together, are agreeable to one another, as those of gold and indigo, and others disagree.

Qu. 15. Are not the species of objects seen with both eyes united where the optic nerves meet before they come into the brain, the fibers on the right side of both nerves uniting there and after union going thence into the brain in the nerve which is on the right side of the head, and the fibers on the left side of both nerves uniting in the same place and after union going into the brain in the nerve which is on the left side of the head, and these two nerves meeting in the brain in such a manner that their fibers make but one entire species or picture, half of which on the right side of the sensorium comes from the right side of both eyes through the right side of both optic nerves to the place where the nerves meet, and from thence on the right side of the head into the brain, and the other half on the left side of the sensorium comes in like manner from the left side of both eyes? For the optic nerves of such animals as look the same way with both eyes (as of men, dogs, sheep, oxen, etc.) meet before they come into the brain, but the optic nerves of such animals as do not look the same way with both eyes (as of fishes and of the chameleon) do not meet, if I am rightly informed.

Qu. 16. When a man in the dark presses either corner of his eye with his finger and turns his eye away from his finger, he will see a circle of colors like those of the feather of a peacock's tail. If the eye and the finger remain quiet, these colors vanish in a second minute of time; but if the finger be moved with a quavering motion, they appear again. Do not these colors arise from such motions excited in the bottom of the eye by the pressure and motion of the finger as, at other times, are excited there by light for causing vision? And do not the motions once excited continue about a second of time before they cease? And when a man, by a stroke upon his eye, sees a flash of light, are not the like motions excited in the retina by the stroke? And when a coal of fire moved nimbly in the circumference of a circle makes the whole circumference appear like a circle of fire, is it not because the motions excited in the bottom of the eye by the rays of light are of a lasting nature and continue till the coal of fire in going round returns to its former place? And considering the lastingness of the motions excited in the bottom of the eye by light, are they not of a vibrating nature?

Qu. 17. If a stone be thrown into stagnating water, the waves excited thereby continue for some time to arise in the place where the stone fell into the water and are propagated from thence in concentric circles upon the surface of the water to great distances. And the vibrations or tremors excited in the air by percussion continue a little time to move from the place of percussion in concentric spheres to great distances. And, in like manner, when a ray of light falls upon the surface of any pellucid body and is there refracted or reflected, may not waves of vibrations or tremors be thereby excited in the refracting or reflecting medium at the point of incidence, and continue to arise there and to be propagated from thence as long as they continue to arise and be propagated, when they are excited in the bottom of the eye by the pressure or motion of the finger or by the light which comes from the coal of fire in the experiments above mentioned? And are not these vibrations propagated from the point of incidence to great distances? And do they not overtake the rays of light, and by overtaking them successively do they not put them into the fits of easy reflection and easy transmission described above? b For if the rays endeavor to recede from the densest part of the vibration, they may be alternately accelerated and retarded by the vibrations overtaking them.

Qu. 18. If in two large, tall cylindrical vessels of glass inverted two little thermometers be suspended so as not to touch the vessels, and the air be drawn out of one of these vessels, and these vessels thus prepared be carried out of a cold place into a warm one, the thermometer *in vacuo* will grow warm as much and almost as soon as the thermometer which is not *in vacuo*. And when the vessels are carried back into the cold place, the thermometer *in vacuo* will grow cold almost as soon as the other thermometer. Is not the heat of the warm room conveyed through the vacuum by the vibrations of a much subtler medium

---

b [This is Newton's famous *theory of fits.* See Note 15, p. 201.]

than air, which, after the air was drawn out, remained in the vacuum? And is not this medium the same with that medium by which light is refracted and reflected, and by whose vibrations light communicates heat to bodies and is put into fits of easy reflection and easy transmission? And do not the vibrations of this medium in hot bodies contribute to the intenseness and duration of their heat? And do not hot bodies communicate their heat to contiguous cold ones by the vibrations of this medium propagated from them into the cold ones? And is not this medium exceedingly more rare and subtle than the air, and exceedingly more elastic and active? And does it not readily pervade all bodies? And is it not (by its elastic force) expanded through all the heavens?

Qu. 19. Does not the refraction of light proceed from the different density of this etherial medium in different places, the light receding always from the denser parts of the medium? And is not the density thereof greater in free and open spaces void of air and other grosser bodies than within the pores of water, glass, crystal, gems, and other compact bodies? For when light passes through glass or crystal, and falling very obliquely upon the farther surface thereof is totally reflected, the total reflection ought to proceed rather from the density and vigor of the medium without and beyond the glass than from the rarity and weakness thereof.

Qu. 20. Does not this etherial medium, in passing out of water, glass, crystal, and other compact and dense bodies into empty spaces, grow denser and denser by degrees, and by that means refract the rays of light, not in a point, but by bending them gradually in curve lines? And does not the gradual condensation of this medium extend to some distance from the bodies, and thereby cause the inflections of the rays of light which pass by the edges of dense bodies at some distance from the bodies?

Qu. 21. Is not this medium much rarer within the dense bodies of the sun, stars, planets, and comets than in the empty celestial spaces between them? And in passing from them to great distances, does it not grow denser and denser perpetually, and thereby cause the gravity of those great bodies toward one another and of their parts toward the bodies, every body endeavoring to go from the denser parts of the medium toward the rarer? For if this medium be rarer within the sun's body than at its surface, and rarer there than at the hundredth part of an inch from its body, and rarer there than at the fiftieth part of an inch from its body, and rarer there than at the orb of Saturn, I see no reason why the increase of density should stop anywhere, and not rather be continued through all distances from the sun to Saturn and beyond. And though this increase of density may at great distances be exceeding slow, yet if the elastic force of this medium be exceedingly great it may suffice to impel bodies from the denser parts of the medium toward the rarer with all that power which we call *gravity*. And that the elastic force of this medium is exceeding great may be gathered from the swiftness of its vibrations. Sounds move about 1,140 English feet in a second minute of time, and in seven or eight minutes of time they move about one hundred English miles. Light moves from the sun to us in about seven or eight minutes of time, which distance is about 70,000,000 English miles, supposing the horizontal parallax of the sun to be about 12 seconds. And the vibrations or pulses of this medium, that they may cause the alternate fits of easy transmission and easy reflection, must be swifter than light, and by consequence above 700,000 times swifter than sounds. And therefore the elastic force of this medium, in proportion to its density, must be above 700,000 x 700,000 (that is, above 490,000,000,000) times greater than the elastic force of the air is in proportion to its density. For the velocities of the pulses of elastic mediums are in a subduplicate ratio of the elasticities and the rarities of the mediums taken together.

As attraction is stronger in small magnets than in great ones in proportion to their bulk, and gravity is greater in the surfaces of small planets than in those of great ones in proportion to their bulk, and small bodies are agitated much more by electric attraction than great ones, so the smallness of the rays of light may contribute very much to the power of the agent by which they are refracted. And so if anyone should suppose that ether (like our air) may contain particles which endeavor to recede from one another (for I do not know what this ether is) and that its particles are exceedingly smaller than those of air or even than those of light, the exceeding smallness of its particles may contribute to the greatness of the force by which those particles may recede from one another, and thereby make that medium exceedingly more rare and elastic than air, and by consequence exceedingly less able to resist the motions of projectiles and exceedingly more able to press upon gross bodies by endeavoring to expand itself.

Qu. 22. May not planets and comets, and all gross bodies, perform their motions more freely and with less resistance in this ethereal medium than in any fluid which fills all space adequately without leaving any pores, and by consequence is much denser than quicksilver or gold? And may not its resistance be so small as to be inconsiderable? For instance, if this *ether* (for so I will call it) should be supposed 700,000 times more elastic than our air, and above 700,000 times more rare, its resistance would be above 600,000,000 times less than that of water. And so small a resistance would scarce make any sensible alteration in the motions of the planets in ten thousand years. If anyone would ask how a medium can be so rare, let him tell me how the air, in the upper parts of the atmosphere, can be above a hundred thousand thousand times rarer than gold. Let him also tell me how an electric body can by friction emit an exhalation so rare and subtle, and yet so potent, as by its emission to cause no sensible diminution of the weight of the electric body, and to be expanded through a sphere whose diameter is above two feet, and yet to be able to agitate and carry up leaf copper or leaf gold at the distance of above a foot from the electric body? And how the effluvia of a magnet can be so rare and subtle as to pass through a plate of glass without any resistance or diminution of their force, and yet so potent as to turn a magnetic needle beyond the glass?

Qu. 23. Is not vision performed chiefly by the vibrations of this medium, excited in the bottom of the eye by the rays of light and propagated through the solid, pellucid, and uniform capillamenta of the optic nerves into the place of sensation? And is not hearing performed by the vibrations either of this or some other medium, excited in the auditory nerves by the tremors of the air and propagated through the solid, pellucid, and uniform capillamenta of those nerves into the place of sensation? And so of the other senses.

Qu. 24. Is not animal motion performed by the vibrations of this medium, excited in the brain by the power of the will and propagated from thence through the solid, pellucid, and uniform capillamenta of the nerves into the muscles, for contracting and dilating them? I suppose that the capillamenta of the nerves are each of them solid and uniform, that the vibrating motion of the ethereal medium may be propagated along them from one end to the other uniformly, and without interruption, for obstructions in the nerves create palsies. And that they may be sufficiently uniform, I suppose them to be pellucid when viewed singly, though the reflections in their cylindrical surfaces may make the whole nerve (composed of many capillamenta) appear opaque and white. For opacity arises from reflecting surfaces such as may disturb and interrupt the motions of this medium.

Qu. 25. Are there not other original properties of the rays of light besides those already described? An instance of another original property we have in the refraction of island crystal, described first by Erasmus Bartholine and afterwards more exactly by Hugenius, in his book *De la Lumière*. This crystal is a pellucid fissile stone, clear as water or crystal of the rock and without color, enduring a red heat without losing its transparency, and in a very strong heat calcining without fusion. Steeped a day or two in water, it loses its natural polish. Being rubbed on cloth, it attracts pieces of straws and other light things, like amber or glass, and with *aqua fortis* it makes an ebullition. It seems to be a sort of talk, and is found in form of an oblique parallelopiped, with six parallelogram sides and eight solid angles. The obtuse angles of the parallelograms are each of them 101 degrees and 52 minutes; the acute ones 78 degrees and 8 minutes. Two of the solid angles opposite to one another, as C and E, are compassed each of them with three of these obtuse angles, and each of the other six with one obtuse and two acute ones. [See diagram on p. 146.] It cleaves easily in planes parallel to any of its sides, and not in any other planes. It cleaves with a glossy polite surface not perfectly plane, but with some little unevenness. It is easily scratched, and by reason of its softness it takes a polish very difficultly. It polishes better upon polished looking glass than upon metal, and perhaps better upon pitch, leather, or parchment. Afterward it must be rubbed with a little oil or white of an egg, to fill up its scratches, whereby it will become very transparent and polite. But for several experiments, it is not necessary to polish it. If a piece of this crystalline stone be laid upon a book, every letter of the book seen through it will appear double, by means of a double refraction. And if any beam of light falls either perpendicularly or in any oblique angle upon any surface of this crystal, it becomes divided into two beams by means of the same double refraction. Which beams are of the same color with the incident beam of light and seem equal to one another in the quantity of their light, or very nearly equal. One of these refractions is performed by the usual rule of optics, the sine of incidence out of air into this crystal being to the sine of refraction as five to three. The other refraction, which may be called the unusual refraction, is performed by the following rule.

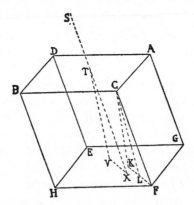

Let ADBC represent the refracting surface of the crystal, C the biggest solid angle at that surface, GEHF the opposite surface, and CK a perpendicular on that surface. This perpendicular makes with the edge of the crystal CF an angle of 19 degrees 3 minutes.

Join KF, and in it take KL, so that the angle KCL be 6 degrees 40 minutes, and the angle LCF 12 degrees 23 minutes. And if ST represent any beam of light incident at T in any angle upon the refracting surface ADBC, let TV be the refracted beam determined by the given portion of the sines 5 to 3, according to the usual rule of optics. Draw VX parallel and equal to KL. Draw it the same way from V in which L lies from K, and joining TX this line TX shall be the other refracted beam carried from T to X, by the unusual refraction.

If therefore the incident beam ST be perpendicular to the refracting surface, the two beams TV and TX, into which it shall become divided, shall be parallel to the lines CK and CL; one of those beams going through the crystal perpendicularly, as it ought to do by the usual laws of optics, and the other, TX, by an unusual refraction diverging from the perpendicular and making with it an angle VTX of about $6\frac{2}{3}$ degrees, as is found by experience. And hence, the plane VTX, and such like planes which are parallel to the plane CFK, may be called the planes of perpendicular refraction. And the coast toward which the lines KL and VX are drawn may be called the coast of unusual refraction.

In like manner crystal of the rock has a double refraction, but the difference of the two refractions is not so great and manifest as in island crystal.

When the beam ST incident on island crystal is divided into two beams, TV and TX, and these two beams arrive at the farther surface of the glass, the beam TV, which was refracted at the first surface after the usual manner, shall be again refracted entirely after the usual manner at the second surface, and the beam TX, which was refracted after the unusual manner in the first surface, shall be again refracted entirely after the unusual manner in the second surface; so that both these beams shall emerge out of the second surface in lines parallel to the first incident beam ST.

And if two pieces of island crystal be placed one after another in such manner that all the surfaces of the latter be parallel to all the corresponding surfaces of the former, the rays which are refracted after the usual manner in the first surface of the first crystal shall be refracted after the usual manner in all the following surfaces, and the rays which are refracted after the unusual manner in the first surface shall be refracted after the unusual manner in all of the following surfaces. And the same thing happens, though the surfaces of the crystals be anyways inclined to one another, provided that their planes of perpendicular refraction be parallel to one another.

And therefore there is an original difference in the rays of light, by means of which some rays are in this experiment constantly refracted after the usual manner and others constantly after the unusual manner; for if the difference be not original, but arises from new modifications impressed on the rays at their first refraction, it would be altered by new modifications in the three following refractions, whereas it suffers no alteration, but is constant and has the same effect upon the rays in all the refractions. The unusual refraction is therefore performed by an original property of the rays. And it remains to be inquired whether the rays have not more original properties than are yet discovered.

Qu. 26. Have not the rays of light several sides, endued with several original properties? For if the planes of perpendicular refraction of the second crystal be at right angles with the planes of perpendicular refraction of the first crystal, the rays which are refracted after the usual manner in passing through the first crystal will be all of them refracted after the unusual manner in passing through the second crystal; and the rays which are refracted after the unusual manner in passing through the first crystal will be all of them refracted after the usual manner in passing through the second crystal. And therefore there are not two sorts of rays differing in their nature from one another, one of which is constantly and in all positions refracted after the usual manner, and the other constantly and in all positions after the unusual manner. The difference between the two sorts of rays in the experiment mentioned in Question 25 was only in the positions of the sides of the rays to the planes of perpendicular refraction. For one and the same ray is here refracted sometimes after the usual and sometimes after the unusual manner, according to the position which its sides have to the crystals. If the sides of the ray are posited the same way to both crystals, it is refracted after the same manner in them both; but if that side of the ray which looks toward the coast of the unusual refraction of the first crystal be 90 degrees from that side of the same ray which looks toward the coast of the unusual refraction of the second crystal (which may be effected by varying the position of the second crystal to the first, and by consequence to the rays of light), the ray shall be refracted after several manners in the several crystals. There is nothing more required to determine whether the rays of light which fall upon the second crystal shall be refracted after the usual or after the unusual manner but to turn about this crystal so that the coast of this crystal's unusual refraction may be on this or on that side of the ray. And therefore every ray may be considered as having four sides or quarters, two of which opposite to one another incline the ray to be refracted after the unusual manner as often as either of them are turned toward the coast of unusual refraction; and the other two, whenever either of them are turned toward the coast of unusual refraction do not incline it to be otherwise refracted than after the usual manner. The two first may therefore be called the sides of unusual refraction. And since these dispositions were in the rays before their incidence on the second, third, and fourth surfaces of the two crystals and suffered no alteration (so far as appears) by the refraction of the rays in their passage through those surfaces, and the rays were refracted by the same laws in all the four surfaces, it appears that those dispositions were in the rays originally and suffered no alteration by the first refraction, and that by means of those dispositions the rays were refracted at their incidence on the first surface of the first crystal, some of them after the usual and some of them after the unusual manner, accordingly as their sides of unusual refraction were then turned toward the coast of the unusual refraction of that crystal or sideways from it.

Every ray of light has therefore two opposite sides originally endued with a property on which the unusual refraction depends, and the other two opposite sides not endued with that property. And it remains to be inquired whether there are not more properties of light by which the sides of the rays differ and are distinguished from one another.[c]

In explaining the difference of the sides of the rays above mentioned, I have supposed that the rays fall perpendicularly on the first crystal. But if they fall obliquely on it, the success is the same. Those rays which are refracted after the usual manner in the first crystal will be refracted after the unusual manner in the second crystal, supposing the planes of perpendicular refraction to be at right angles with one another, as above; and on the contrary.

If the planes of the perpendicular refraction of the two crystals be neither parallel nor perpendicular to one another, but contain an acute angle, the two beams of light which emerge out of the first crystal will be each of them divided into two more at their incidence on the second crystal. For in this case the rays in each of the two beams will some of them have their sides of unusual refraction, and some of them their other sides turned toward the coast of the unusual refraction of the second crystal.

Qu. 27. Are not all hypotheses erroneous which have hitherto been invented for explaining the phenomena of light by new modifications of the rays? For those phenomena depend not upon new modifications, as has been supposed, but upon the original and unchangeable properties of the rays.

Qu. 28. Are not all hypotheses erroneous in which light is supposed to consist in pression or motion, propagated through a fluid medium? For in all these hypotheses the phenomena of light have been hitherto explained by supposing that they arise from new modifications of the rays, which is an erroneous supposition.

If light consisted only in pression propagated without actual motion, it would not be able to agitate and heat the bodies which refract and reflect it. If it consisted in motion propagated to all distances in an instant, it would require an infinite force every moment, in every shining particle, to generate that motion. And if it consisted in pression or motion, propagated either in an instant or in time, it would bend into the shadow. For pression or motion

---

[c] [Here Newton has made the discovery of the *polarization of light.* The name 'polarization' is a result of Newton's reference to the poles of magnets. See Query 29.]

cannot be propagated in a fluid in right lines beyond an obstacle which stops part of the motion, but will bend and spread every way into the quiescent medium which lies beyond the obstacle. Gravity tends downward, but the pressure of water arising from gravity tends every way with equal force, and is propagated as readily and with as much force sideways as downward, and through crooked passages as through straight ones. The waves on the surface of stagnating water, passing by the sides of a broad obstacle which stops part of them, bend afterward and dilate themselves gradually into the quiet water behind the obstacle. The waves, pulses, or vibrations of the air, wherein sounds consist, bend manifestly, though not so much as the waves of water. For a bell or a cannon may be heard beyond a hill which intercepts the sight of the sounding body, and sounds are propagated as readily through crooked pipes as through straight ones. But light is never known to follow crooked passages nor to bend into the shadow. For the fixed stars by the interposition of any of the planets cease to be seen. And so do the parts of the sun by the interposition of the moon, Mercury, or Venus. The rays which pass very near to the edges of any body are bent a little by the action of the body, as we showed above; but this bending is not toward but from the shadow, and is performed only in the passage of the ray by the body and at a very small distance from it. So soon as the ray is past the body, it goes right on.

To explain the unusual refraction of island crystal by pression or motion propagated has not hitherto been attempted (to my knowledge) except by Huygens, who for that end supposed two several vibrating mediums within that crystal. But when he tried the refractions in two successive pieces of that crystal and found them such as is mentioned above, he confessed himself at a loss for explaining them. For pressions or motions propagated from a shining body through a uniform medium must be on all sides alike, whereas by those experiments it appears that the rays of light have different properties in their different sides. He suspected that the pulses of ether in passing through the first crystal might receive certain new modifications, which might determine them to be propagated in this or that medium within the second crystal, according to the position of that crystal. But what modifications those might be he could not say, nor think of anything satisfactory in that point.[d] And if he had known that the unusual refraction depends not on new modifications, but on the original and unchangeable dispositions of the rays, he would have found it as difficult to explain how those dispositions which he supposed to be impressed on the rays by the first crystal could be in them before their incidence on that crystal, and in general how all rays emitted by shining bodies can have those dispositions in them from the beginning. To me, at least, this seems inexplicable, if light be nothing else than pression or motion propagated through ether.

And it is as difficult to explain by these hypotheses how rays can be alternately in fits of easy reflection and easy transmission, unless perhaps one might suppose that there are in all space two ethereal vibrating mediums, and that the vibrations of one of them constitute light and the vibrations of the other are swifter, and as often as they overtake the vibrations of the first put them into those fits. But how two ethers can be diffused through all space, one of which acts upon the other and by consequence is reacted upon, without retarding, shattering, dispersing, and confounding one another's motions, is inconceivable. And against filling the heavens with fluid mediums, unless they be exceeding rare, a great objection arises from the regular and very lasting motions of the planets and comets in all manner of courses through the heavens. For thence it is manifest that the heavens are void of all sensible resistance, and by consequence of all sensible matter.

For the resisting power of fluid mediums arises partly from the attrition of the parts of the medium and partly from the *vis inertiae* of the matter. That part of the resistance of a spherical body which arises from the attrition of the parts of the medium is very nearly as the diameter, or at the most as the *factum* of the diameter, and the velocity of the spherical body together. And that part of the resistance which arises from the *vis inertiae* of the matter is as the square of that *factum*. And by this differ-

---

d "Mais pour dire comment cela se fait, je n'ay rien trove jusqu'ici qui me satisfasse." C. H., *De la Lumière,* c. 5, p. 91.

ence the two sorts of resistance may be distinguished from one another in any medium, and these being distinguished it will be found that almost all the resistance of bodies of a competent magnitude moving in air, water, quicksilver, and suchlike fluids with a competent velocity arises from the *vis inertiae* of the parts of the fluid.

Now that part of the resisting power of any medium which arises from the tenacity, friction, or attrition of the parts of the medium may be diminished by dividing the matter into smaller parts and making the parts more smooth and slippery; but that part of the resistance which arises from the *vis inertiae* is proportional to the density of the matter and cannot be diminished by dividing the matter into smaller parts, nor by any other means than by decreasing the density of the medium. And for these reasons the density of fluid mediums is very nearly proportional to their resistance. Liquors which differ not much in density, as water, spirit of wine, spirit of turpentine, hot oil, differ not much in resistance. Water is thirteen or fourteen times lighter than quicksilver, and by consequence thirteen or fourteen times rarer; and its resistance is less than that of quicksilver in the same proportion, or thereabouts, as I have found by experiments made with pendulums. The open air in which we breathe is eight or nine hundred times lighter than water, and by consequence eight or nine hundred times rarer; and accordingly its resistance is less than that of water in the same proportion, or thereabouts, as I have also found by experiments made with pendulums. And in thinner air the resistance is still less, and at length, by rarifying the air, becomes insensible. For small feathers falling in the open air meet with great resistance, but in a tall glass well emptied of air they fall as fast as lead or gold, as I have seen tried several times. Whence the resistance seems still to decrease in proportion to the density of the fluid. For I do not find by any experiments that bodies moving in quicksilver, water, or air meet with any other sensible resistance than what arises from the density and tenacity of those sensible fluids, as they would do if the pores of those fluids and all other spaces were filled with a dense and subtle fluid. Now if the resistance in a vessel well emptied of air was but a hundred times less than in the open air, it would be about a million of times less than in quicksilver. But it seems to be much less in such a vessel, and still much less in the heavens, at the height of three or four hundred miles from the earth, or above. For Mr. Boyle has showed that air may be rarified above ten thousand times in vessels of glass, and the heavens are much emptier of air than any vacuum we can make below. For since the air is compressed by the weight of the incumbent atmosphere and the density of air is proportional to the force compressing it, it follows by computation that, at the height of about seven and a half English miles from the earth, the air is four times rarer than at the surface of the earth; and at the height of 15 miles it is sixteen times rarer than that at the surface of the earth; and at the height of $22\frac{1}{2}$, 30, or 38 miles it is respectively 64, 256, or 1,024 times rarer, or thereabouts; and at the height of 76, 152, 228 miles, it is about 1,000,000, 1,000,000,000,000, or 1,000,000,-000,000,000,000 times rarer, and so on.

Heat promotes fluidity very much by diminishing the tenacity of bodies. It makes many bodies fluid which are not fluid in cold and increases the fluidity of tenacious liquids, as of oil, balsam, and honey, and thereby decreases their resistance. But it decreases not the resistance of water considerably, as it would do if any considerable part of the resistance of water arose from the attrition or tenacity of its parts. And therefore the resistance of water arises principally and almost entirely from the *vis inertiae* of its matter; and by consequence, if the heavens were as dense as water they would not have much less resistance than water, if as dense as quicksilver they would not have much less resistance than quicksilver, if absolutely dense or full of matter without any vacuum, let the matter be never so subtle and fluid, they would have a greater resistance than quicksilver. A solid globe in such a medium would lose above half its motion in moving three times the length of its diameter, and a globe not solid (such as are the planets) would be retarded sooner. And therefore to make way for the regular and lasting motions of the planets and comets, it is necessary to empty the heavens of all matter, except perhaps some very thin vapors, steams, or effluvia arising from the atmospheres of the earth, planets, and comets, and from such an exceedingly rare ethereal medium as we described above. A dense fluid can be of no use for explaining the phenomena of nature, the motions of the planets and comets being better explained without it. It serves only to disturb and retard the motions of those great bodies and make the frame of nature languish; and in the pores of bodies it serves only to stop the vibrating motions of their parts, wherein their heat and activity consists. And as it is of no use, and hinders the operations of nature and makes her languish, so there is no evidence for its existence, and therefore it ought to be rejected. And if it be rejected, the hypotheses that light consists in pression or motion propagated through such a medium are rejected with it.

And for rejecting such a medium we have the authority of those the oldest and most celebrated philosophers of Greece and Phoenicia, who made a vacuum and atoms and the gravity of atoms the first principles of their philosophy, tacitly attributing gravity to some other cause than dense matter. Later philosophers banish the consideration of such a cause out of natural philosophy, feigning hypotheses for explaining all things mechanically and referring other causes to metaphysics, whereas the main business of natural philosophy is to argue from phenomena without feigning hypotheses and to deduce causes from effects, till we come to the very first cause, which certainly is not mechanical; and not only to unfold the mechanism of the world, but chiefly to resolve these and suchlike questions. What is there in places almost empty of matter, and whence is it that the sun and planets gravitate toward one another, without dense matter between them? Whence is it that nature does nothing in vain, and whence arises all that order and beauty which we see in the world? To what end are comets, and whence is it that planets move all one and the same way in orbs concentric while comets move all manner of ways in orbs very eccentric, and what hinders the fixed stars from falling upon one another? How came the bodies of animals to be contrived with so much art, and for what ends were their several parts? Was the eye contrived without skill in optics and the ear without knowledge of sounds? How do the motions of the body follow from the will, and whence is the instinct in animals? Is not the sensory of animals that place to which the sensitive substance is present and into which the sensible species of things are carried through the nerves and brain, that there they may be perceived by their immediate presence to that substance? And these things being rightly dispatched, does it not appear from phenomena that there is a Being, incorporeal, living, intelligent, omnipresent, who in infinite space, as it were in his sensory, sees the things themselves intimately and thoroughly perceives them, and comprehends them wholly by their immediate presence to himself, of which things the images only carried through the organs of sense into our little sensoriums are there seen and beheld by that which in us perceives and thinks? And though every true step made in this philosophy brings us not immediately to the knowledge of the first cause, yet it brings us nearer to it, and on that account is to be highly valued.

Qu. 29. Are not the rays of light very small bodies emitted from shining substances? For such bodies will pass through uniform mediums in right lines without bending into the shadow, which is the nature of the rays of light. They will also be capable of several properties and be able to conserve their properties unchanged in passing through several mediums, which is another condition of the rays of light. Pellucid substances act upon the rays of light at a distance in refracting, reflecting, and inflecting them, and the rays mutually agitate the parts of those substances at a distance for heating them; and this action and reaction at a distance very much resembles an attractive force between bodies. If refraction be performed by attraction of the rays, the sines of incidence must be to the sines of refraction in a given proportion, as we showed in our *Principles of Philosophy,* and this rule is true by experience. The rays of light, in going out of glass into a vacuum, are bent toward the glass, and if they fall too obliquely on the vacuum they are bent backward into the glass and totally reflected; and this reflection cannot be ascribed to the resistance of an absolute vacuum, but must be caused by the power of the glass attracting the rays at their going out of it into the vacuum and bringing them back. For if the farther surface of the glass be moistened with water or clear oil, or liquid and clear honey, the rays which would otherwise be reflected will go into the water, oil, or honey, and therefore are not reflected before they arrive at the farther surface of the glass and begin to go out of it. If they go out of it into the water, oil, or honey, they go on, because the attraction of the glass is almost balanced and rendered ineffectual by the contrary attraction of the liquor. But if they go out of it into a vacuum which has no attraction to balance that of the glass, the attraction of the glass either bends and refracts them or brings them back and reflects them. And this is still more evident by laying together two prisms of glass or two object glasses of very long telescopes, the one plane, the other a little convex, and so compressing them that they do not fully touch, nor are too far asunder. For the light which falls upon the farther surface of the first glass where the interval between the glasses is not above the ten-hundred-thousandth part of an inch will go through that surface, and through the air or vacuum between the glasses, and enter into the second glass, as was explained in the first, fourth, and eighth observations of the first part of the Second Book. But if the second glass be taken away, the light which goes out of the second surface of the first glass into the air or vacuum will not go on forward, but turns back into the first glass and is reflected; and therefore it is drawn back by the power of the first glass, there being nothing else to turn it back. Nothing more is requisite for producing all the variety of colors and degrees of refrangibility than that the rays of light be bodies of different sizes, the least of which may take violet, the weakest and darkest of the colors, and be more easily diverted by refracting surfaces from the right course; and the rest, as they are bigger and bigger, may make the stronger and more lucid colors, blue, green, yellow, and red, and be more and more difficultly diverted. Nothing more is requisite for putting the rays of light into fits of easy reflection and easy transmission than that they be small bodies which, by their attractive powers or some other force, stir up vibrations in what they act upon, which vibrations being swifter than the rays overtake them successively and agitate them so as by turns to increase and decrease their velocities and thereby put them into those fits. And lastly, the unusual refraction of island crystal looks very much as if it were performed by some kind of attractive virtue lodged in certain sides both of the rays and of the particles of the crystal. For were it not for some kind of disposition or virtue lodged in some sides of the particles of the crystal, and not in their other sides, and which inclines and bends the rays toward the coast of unusual refraction, the rays which fall perpendicularly on the crystal would not be refracted toward that coast rather than toward any other coast, both at their incidence and at their emergence, so as to emerge perpendicularly by a contrary situation of the coast of unusual refraction at the second surface, the crystal acting upon the rays after they have passed through it and are emerging into the air or, if you please, into a vacuum. And since the crystal by this disposition or virtue does not act upon the rays, unless when one of their sides of unusual refraction looks toward that coast, this argues a virtue or disposition in those sides of the rays which answers to and sympathizes with that virtue or disposition of the crystal as the poles of two magnets answer to one another. And as magnetism may be intended and remitted, and is found only in the magnet and in iron, so this virtue of refracting the perpendicular rays is greater in island crystal, less in crystal of the rock, and is not yet found in other bodies. I do not say that this virtue is magnetical; it seems to be of another kind. I only say that whatever it be, it is difficult to conceive how the rays of light, unless they be bodies, can have a permanent virtue in two of their sides which is not in their other sides, and this without any regard to their position to the space or medium through which they pass.

What I mean in this question by a vacuum and by the attractions of the rays of light toward glass or crystal may be understood by what was said in Questions 18, 19, and 20.

Quest. 30.[e] Are not gross bodies and light convertible into one another, and may not bodies receive much of their activity from the particles of light which enter their composition? For all fixed bodies being heated emit light so long as they continue sufficiently

---

[e] [Here and in Quest. 31 we have a picture of some of Newton's interests in chemistry. He begins by advocating the possibility of the transmutation of metals, and even the convertibility of gross bodies and light. See also Boyle's *Works,* Vol. I, pp. cv, cxxx.]

hot, and light mutually stops in bodies as often as its rays strike upon their parts, as we showed above. I know no body less apt to shine than water; and yet water by frequent distillations changes into fixed earth, as Mr. Boyle has tried, and then this earth being enabled to endure a sufficient heat shines by heat like other bodies.

The changing of bodies into light and light into bodies is very conformable to the course of nature, which seems delighted with transmutations. Water, which is a very fluid, tasteless salt, she changes by heat into vapor, which is a sort of air, and by cold into ice, which is a hard, pellucid, brittle, fusible stone; and this stone returns into water by heat, and vapor returns into water by cold. Earth by heat becomes fire and by cold returns into earth. Dense bodies by fermentation rarify into several sorts of air, and this air by fermentation, and sometimes without it, returns into dense bodies. Mercury appears sometimes in the form of a fluid metal; sometimes in the form of a hard, brittle metal; sometimes in the form of a corrosive, pellucid salt called *sublimate;* sometimes in the form of a tasteless, pellucid, volatile white earth called *mercurius dulcis,* or in that of a red opaque volatile earth called cinnabar, or in that of a red or white precipitate, or in that of a fluid salt; and in distillation it turns into vapor, and being agitated *in vacuo* it shines like fire. And after all these changes it returns again into its first form of mercury. Eggs grow from insensible magnitudes and change into animals, tadpoles into frogs, and worms into flies. All birds, beasts, and fishes, insects, trees, and other vegetables, with their several parts, grow out of water and watery tinctures and salts, and by putrefaction return again into watery substances. And water standing a few days in the open air yields a tincture which (like that of malt), by standing longer, yields a sediment and a spirit, but before putrefaction is fit nourishment for animals and vegetables. And among such various and strange transmutations, why may not nature change bodies into light and light into bodies?

Quest. 31. Have not the small particles of bodies certain powers, virtues, or forces by which they act at a distance, not only upon the rays of light for reflecting, refracting, and inflecting them, but also upon one another for producing a great part of the phenomena of nature? For it is well known that bodies act one upon another by the attractions of gravity, magnetism, and electricity; and these instances show the tenor and course of nature, and make it not improbable but that there may be more attractive powers than these. For nature is very consonant and conformable to herself. How these attractions may be performed, I do not here consider. What I call 'attraction' may be performed by impulse, or by some other means unknown to me. I use that word here to signify only in general any force by which bodies tend toward one another, whatsoever be the cause. For we must learn from the phenomena of nature what bodies attract one another, and what are the laws and properties of the attraction, before we inquire the cause by which the attraction is performed. The attractions of gravity, magnetism, and electricity reach to very sensible distances, and so have been observed by vulgar eyes; and there may be others which reach to so small distances as hitherto escape observation, and perhaps electrical attraction may reach to such small distances even without being excited by friction.

For when salt of tartar runs *per deliquium*, is not this done by an attraction between the particles of the salt of tartar and the particles of the water which float in the air in the form of vapors? And why does not common salt, or saltpeter, or vitriol run *per deliquium*, but for want of such an attraction? Or why does not salt of tartar draw more water out of the air than in a certain proportion to its quantity, but for want of an attractive force after it is satiated with water? And whence is it but from this attractive power that water which alone distils with a gentle lukewarm heat will not distil from salt of tartar without a great heat? And is it not from the like attractive power between the particles of oil of vitriol and the particles of water that oil of vitriol draws to it a good quantity of water out of the air, and after it is satiated draws no more, and in distillation lets go the water very difficultly? And when water and oil of vitriol poured successively into the same vessel grow very hot in the mixing, does not this heat argue a great motion in the parts of the liquors? And does not this motion argue that the parts of the two liquors in mixing coalesce with violence, and by consequence rush toward one another with an accelerated motion? And when *aqua fortis,* or spirit of vitriol, poured upon filings of iron dissolves the filings with a great heat and ebullition, is not this heat and ebullition effected by a violent motion of the parts; and does not that motion argue that the acid parts of the liquor rush toward the parts of the metal with violence and run forcibly into its pores till they get between its outmost particles and the main mass of the metal, and surrounding those particles loosen them from the main mass and set them at liberty to float off into the water? And when the acid particles, which alone would distil with an easy heat, will not separate from the particles of the metal without a very violent heat, does not this confirm the attraction between them?

When spirit of vitriol poured upon common salt or saltpeter makes an ebullition with the salt and unites with it, and in distillation the spirit of the common salt or saltpeter comes over much easier than it would do before and the acid part of the spirit of vitriol stays behind, does not this argue that the fixed alkali of the salt attracts the acid spirit of the vitriol more strongly than its own spirit and, not being able to hold them both, lets go its own? And when oil of vitriol is drawn off from its weight of niter and from both the ingredients a compound spirit of niter is distilled, and two parts of this spirit are poured on one part of oil of cloves or caraway seeds, or of any ponderous oil of vegetable or animal substances, or oil of turpentine thickened with a little balsam of sulphur, and the liquors grow so very hot in mixing as presently to send up a burning flame, does not this very great and sudden heat argue that the two liquors mix with violence, and that their parts in mixing run toward one another with an accelerated motion and clash with the greatest force? And is it not for the same reason that well-rectified spirit of wine poured on the same compound spirit flashes; and that the *pulvis fulminans,* composed of sulphur, niter, and salt of tartar, goes off with a more sudden and violent explosion than gunpowder, the acid spirits of the sulphur and niter rushing toward one another and toward the salt of tartar with so great a violence as by the shock to turn the whole at once into vapor and flame? Where the dissolution is slow, it makes a slow ebullition and a gentle heat; and where it is quicker, it makes a greater ebullition with more heat; and where it is done at once, the ebullition is contracted into a sudden blast or violent explosion, with a heat equal to that of fire and flame. So when a dram of the above-mentioned compound spirit of niter was poured upon half a dram of oil of caraway seeds *in vacuo,* the mixture immediately made a flash like gunpowder and burst the exhausted receiver, which was a glass six inches wide and eight inches deep. And even the gross body of sulphur powdered, and with an equal weight of iron filings and a little water made into paste, acts upon the iron, and in five or six hours grows too hot to be touched and emits a flame. And by these experiments compared with the great quantity of sulphur with which the earth abounds, and the warmth of the interior parts of the earth and hot springs and burning mountains, and with damps, mineral coruscations, earthquakes, hot suffocating exhalations, hurricanes, and spouts, we may learn that sulphureous steams abound in the bowels of the earth and ferment with minerals, and sometimes take fire with a sudden coruscation and explosion, and if pent up in subterraneous caverns burst the caverns with a great shaking of the earth as in springing of a mine. And then the vapor generated by the explosion, expiring through the pores of the earth, feels hot and suffocates, and makes tempests and hurricanes, and sometimes causes the land to slide or the sea to boil, and carries up the water thereof in drops, which by their weight fall down again in spouts. Also some sulphureous steams, at all times when the earth is dry, ascending into the air, ferment there with nitrous acids and sometimes, taking fire, cause lightning and thunder and fiery meteors. For the air abounds with acid vapors fit to promote fermentations, as appears by the rusting of iron and copper in it, the kindling of fire by blowing, and the beating of the heart by means of respiration. Now the above-mentioned motions are so great and violent as to show that in fermentations the particles of bodies which almost rest are put into new motions by a very potent principle, which acts upon them only when they approach one another, and causes them to meet and clash with great violence and grow hot with the motion, and dash one another into pieces and vanish into air, and vapor, and flame.

When salt of tartar *per deliquium,* being poured into the solution of any metal, precipitates the metal and makes it fall down to the bottom of the liquor in the form of mud, does not this argue that the acid particles are attracted more strongly by the salt of tartar than by the metal, and by the stronger attraction go from the metal to the salt of tartar? And so when a solution of iron in *aqua fortis* dissolves the *lapis calaminaris* and lets go the iron, or a solution of copper dissolves iron immersed in it and lets go the copper, or a solution of silver dissolves copper and lets go the silver, or a solution of mercury in *aqua fortis* being poured upon iron, copper, tin, or lead dissolves the metal and lets go the mercury, does not this argue that the acid particles of the *aqua fortis* are attracted more strongly by the *lapis calaminaris* than by iron, and more strongly by iron than by copper, and more strongly by copper than by silver, and more strongly by iron, copper, tin, and lead than by mercury? And is it not for the same reason that iron requires more *aqua fortis* to dissolve it than copper and copper more than the other metals, and that of all metals iron is dissolved most easily and is most apt to rust, and next after iron copper?

When oil of vitriol is mixed with a little water, or is run *per deliquium,* and in distillation the water ascends difficultly and brings over with it some part of the oil of vitriol in the form of spirit of vitriol, and this spirit being poured upon iron, copper, or salt of tartar unites with the body and lets go the water, does not this show that the acid spirit is attracted by the water and more attracted by the fixed body than by the water, and therefore lets go the water to close with the fixed body? And is it not for the same reason that the water and acid spirits which are mixed together in vinegar, *aqua fortis,* and spirit of salt cohere and rise together in distillation; but if the *menstruum* be poured on salt of tartar, or on lead or iron or any fixed body which it can dissolve, the acid by a stronger attraction adheres to the body and lets go the water? And is it not also from a mutual attraction that the spirits of soot and sea salt unite and compose the particles of sal armoniac, which are less volatile than before, because grosser and freer from water; and that the particles of sal armoniac in sublimation carry up the particles of antimony, which will not sublime alone; and that the particles of mercury uniting with the acid particles of spirit of salt compose mercury sublimate, and with the particles of sulphur compose cinnabar; and that the particles of spirit of wine and spirit of urine well rectified unite and, letting go the water which dissolved them, compose a consistent body; and that in subliming cinnabar from salt of tartar or from quicklime, the sulphur by a stronger attraction of the salt or lime lets go the mercury and stays with the fixed body; and that when mercury sublimate is sublimed from antimony or from regulus of antimony, the spirit of salt lets go the mercury and unites with the antimonial metal which attracts it more strongly, and stays with it till the heat be great enough to make them both ascend together, and then carries up the metal with it in the form of a very fusible salt, called 'butter of antimony,' although the spirit of salt alone be almost as volatile as water and the antimony alone as fixed as lead?

When *aqua fortis* dissolves silver and not gold, and *aqua regia* dissolves gold and not silver, may it not be said that *aqua fortis* is subtle enough to penetrate gold as well as silver, but wants the attractive force to give it entrance; and that *aqua regia* is subtle enough to penetrate silver as well as gold, but wants the attractive force to give it entrance? For *aqua regia* is nothing else than *aqua fortis* mixed with some spirit of salt, or with sal armoniac; and even common salt dissolved in *aqua fortis* enables the *menstruum* to dissolve gold, though the salt be a gross body. When, therefore, spirit of salt precipitates silver out of *aqua fortis,* is it not done by attracting and mixing with the *aqua fortis,* and not attracting or perhaps repelling silver? And when water precipitates antimony out of the sublimate of antimony and sal armoniac, or out of butter of antimony, is it not done by its dissolving, mixing with, and weakening the sal armoniac or spirit of salt, and its not attracting or perhaps repelling the antimony? And is it not for want of an attractive virtue between the parts of water and oil, of quicksilver and antimony, of lead and iron, that these substances do not mix; and by a weak attraction that quicksilver and copper mix difficulty; and from a strong one that quicksilver and tin, antimony and iron, water and salts, mix readily? And in general, is it not from the same principle that heat congregates homogeneal bodies and separates heterogeneal ones?

When arsenic with soap gives a regulus, and with mercury sublimate a volatile fusible salt like butter of antimony, does not this show that arsenic, which is a substance totally volatile, is compounded of fixed and volatile parts, strongly cohering by a mutual attraction, so that the volatile will not ascend without carrying up the fixed? And so when an equal weight of spirit of wine and oil of vitriol are digested together, and in distillation yield two fragrant and volatile spirits which will not mix with one another, and a fixed black earth remains behind, does not this show that oil of vitriol is composed of volatile and fixed parts strongly united by attraction, so as to ascend together in form of a volatile, acid, fluid salt, until the spirit of wine attracts and separates the volatile parts from the fixed? And therefore, since oil of sulphur *per campanam* is of the same nature with oil of vitriol, may it not be inferred that sulphur is also a mixture of volatile and fixed parts so strongly cohering by attraction as to ascend together in sublimation? By dissolving flowers of sulphur in oil of turpentine and distilling the solution, it is found that sulphur is composed of an inflammable thick oil or fat bitumen, an acid salt, a very fixed earth, and a little metal. The three first were found not much unequal to one another, the fourth in so small a quantity as scarce to be worth considering. The acid salt dissolved in water is the same with oil of sulphur *per campanam,* and abounding much in the bowels of the earth, and particularly in markasites, unites itself to the other ingredients of the markasite, which are bitumen, iron, copper, and earth, and with them compounds allum, vitriol, and sulphur. With the earth alone it compounds allum; with the metal alone, or metal and earth together, it compounds vitriol; and with the bitumen and earth it compounds sulphur. Whence it comes to pass that markasites abound with those three minerals. And is it not from the mutual attraction of the ingredients that they stick together for compounding these minerals and that the bitumen carries up the other ingredients of the sulphur, which without it would not sublime? And the same question may be put concerning all, or almost all the gross bodies in nature. For all the parts of animals and vegetables are composed of substances volatile and fixed, fluid and solid, as appears by their analysis; and so are salts and minerals, so far as chemists have been hitherto able to examine their composition.

When mercury sublimate is resublimed with fresh mercury and becomes *mercurius dulcis,* which is a white, tasteless earth scarce dissolvable in water, and *mercurius dulcis* resublimed with spirit of salt returns into mercury sublimate; and when metals corroded with a little acid turn into rust, which is an earth tasteless and indissolvable in water, and this earth imbibed with more acid becomes a metallic salt; and when some stones, as spar of lead, dissolved in proper *menstruums* become salts: do not these things show that salts are dry earth and watery acid united by attraction, and that the earth will not become a salt without so much acid as makes it dissolvable in water? Do not the sharp and pungent tastes of acids arise from the strong attraction whereby the acid particles rush upon and agitate the particles of the tongue? And when metals are dissolved in acid *menstruums* and the acids in conjunction with the metal act after a different manner, so that the compound has a different taste much milder than before, and sometimes a sweet one, is it not because the acids adhere to the metallic particles and thereby lose much of their activity? And if the acid be in too small a proportion to make the compound dissolvable in water, will it not by adhering strongly to the metal become inactive and lose its taste, and the compound be a tasteless earth? For such things as are not dissolvable by the moisture of the tongue act not upon the taste.

As gravity makes the sea flow round the denser and weightier parts of the globe of the earth, so the attraction may make the watery acid flow round the denser and compacter particles of earth for composing the particles of salt. For otherwise the acid would not do the office of a medium between the earth and common water for making salts dissolvable in the water; nor would salt of tartar readily draw off the acid from dissolved metals, nor metals the acid from mercury. Now as in the great globe of the earth and sea the densest bodies by their gravity sink down in water and always endeavor to go toward the center of the globe, so in particles of salt the densest matter may always endeavor to approach the center of the particle; so that a particle of salt may be compared to a chaos, being dense, hard, dry, and earthy in the center and rare, soft, moist, and watery in the circumference. And hence it seems to be that salts are of a lasting nature, being scarce destroyed, unless by drawing away their watery parts by violence or by letting them soak into the pores of the central earth by a gentle heat in putrefaction, until the earth be dissolved by the water and separated into smaller particles, which by reason of their smallness make the rotten compound appear of a black color. Hence also it may be that the parts of animals and vegetables preserve their several forms and assimilate their nourishment, the soft and moist nourishment easily changing its texture by a gentle heat and motion till it becomes like the dense, hard, dry, and durable earth in the center of each particle. But when the nourishment grows unfit to be assimilated or the central earth grows too feeble to assimilate it, the motion ends in confusion, putrefaction, and death.

If a very small quantity of any salt or vitriol be dissolved in a great quantity of water, the particles of the salt or vitriol will not sink to the bottom, though they be heavier in specie than the water, but will evenly diffuse themselves into all the water, so as to make it as saline at the top as at the bottom. And does not this imply that the parts of the salt or vitriol recede from one another, and endeavor to expand themselves and get as far asunder as the quantity of water in which they float will allow? And does not this endeavor imply that they have a repulsive force by which they fly from one another, or at least that they attract the water more strongly than they do one another? For as all things ascend in water which are less attracted than water by the gravitating power of the earth, so all the particles of salt which float in water and are less attracted than water by any one particle of salt must recede from that particle and give way to the more attracted water.

When any saline liquor is evaporated to a cuticle and let cool, the salt concretes in regular figures; which argues that the particles of the salt, before they concreted, floated in the liquor at equal distances in rank and file, and by consequence that they acted upon one another by some power which at equal distances is equal, at unequal distances unequal. For by such a power they will range themselves uniformly, and without it they will float irregularly and come together as irregularly. And since the particles of island crystal act all the same way upon the rays of light for causing the unusual refraction, may it not be supposed that in the formation of this crystal the particles not only ranged themselves in rank and file for concreting in regular figures, but also by some kind of polar virtue turned their homogeneal sides the same way?

The parts of all homogeneal hard bodies which fully touch one another stick together very strongly. And for explaining how this may be, some have invented hooked atoms,[f] which is begging the question; and others tell us that bodies are glued together by rest,[g] that is, by an occult quality, or rather by nothing; and others, that they stick together by conspiring motions, that is, by relative rest amongst themselves. I had rather infer from their cohesion that their particles attract one another by some force, which in immediate contact is exceeding strong, at small distances performs the chemical operations above mentioned, and reaches not far from the particles with any sensible effect.

All bodies seem to be composed of hard particles; for otherwise fluids would not congeal, as water, oils, vinegar, and spirit or oil of vitriol do by freezing, mercury by fumes of lead, spirit of niter and mercury by dissolving the mercury and evaporating the flegm, spirit of wine and spirit of urine by deflegming and mixing them, and spirit of urine and spirit of salt by subliming them together to make sal armoniac. Even the rays of light seem to be hard bodies, for otherwise they would not retain different properties in their different sides. And therefore hardness may be reckoned the property of all uncompounded matter. At least this seems to be as evident as the universal impenetrability of matter. For all bodies, so far as experience reaches, are either hard or may be

f [Democritus.]
g [Descartes.]

hardened, and we have no other evidence of universal impenetrability besides a large experience without an experimental exception. Now if compound bodies are so very hard as we find some of them to be, and yet are very porous and consist of parts which are only laid together, the simple particles which are void of pores and were never yet divided must be much harder. For such hard particles being heaped up together can scarce touch one another in more than a few points, and therefore must be separable by much less force than is requisite to break a solid particle, whose parts touch in all the space between them, without any pores or interstices to weaken their cohesion. And how such very hard particles which are only laid together and touch only in a few points can stick together, and that so firmly as they do, without the assistance of something which causes them to be attracted or pressed toward one another, is very difficult to conceive.

The same thing I infer also from the cohering of two polished marbles *in vacuo* and from the standing of quicksilver in the barometer at the height of 50, 60, or 70 inches, or above, whenever it is well purged of air and carefully poured in, so that its parts be everywhere contiguous both to one another and to the glass. The atmosphere by its weight presses the quicksilver into the glass, to the height of 29 or 30 inches. And some other agent raises it higher, not by pressing it into the glass, but by making its parts stick to the glass and to one another. For upon any discontinuation of parts, made either by bubbles or by shaking the glass, the whole mercury falls down to the height of 29 or 30 inches.

And of the same kind with these experiments are those that follow. If two plane polished plates of glass (suppose two pieces of a polished looking glass) be laid together so that their sides be parallel and at a very small distance from one another, and then their lower edges be dipped into water, the water will rise up between them. And the less the distance of the glasses is, the greater will be the height to which the water will rise. If the distance be about the hundredth part of an inch, the water will rise to the height of about an inch; and if the distance be greater or less in any proportion, the height will be reciprocally proportional to the distance very nearly. For the attractive force of the glasses is the same, whether the distance between them be greater or less; and the weight of the water drawn up is the same, if the height of it be reciprocally proportional to the distance of the glasses. And in like manner water ascends between two marbles polished plane, when their polished sides are parallel and at a very little distance from one another. And if slender pipes of glass be dipped at one end into stagnating water, the water will rise up within the pipe; and the height to which it rises will be reciprocally proportional to the diameter of the cavity of the pipe and will equal the height to which it rises between two planes of glass, if the semidiameter of the cavity of the pipe be equal to the distance between the planes, or thereabouts. And these experiments succeed after the same manner *in vacuo* as in the open air (as has been tried before the Royal Society), and therefore are not influenced by the weight or pressure of the atmosphere.

And if a large pipe of glass be filled with sifted ashes well pressed together in the glass, and one end of the pipe be dipped into stagnating water, the water will rise up slowly in the ashes, so as in the space of a week or fortnight to reach up within the glass to the height of 30 or 40 inches above the stagnating water. And the water rises up to this height by the action only of those particles of the ashes which are upon the surface of the elevated water; the particles which are within the water attracting or repelling it as much downward as upward. And therefore the action of the particles is very strong. But the particles of the ashes being not so dense and close together as those of glass, their action is not so strong as that of glass, which keeps quicksilver suspended to the height of 60 or 70 inches, and therefore acts with a force which would keep water suspended to the height of above 60 feet.

By the same principle, a sponge sucks in water and the glands in the bodies of animals, according to their several natures and dispositions, suck in various juices from the blood.

If two plane polished plates of glass, three or four inches broad and twenty or twenty-five long, be laid one of them parallel to the horizon, the other upon the first, so as at one of their ends to touch one another and contain an angle of about 10 or 15 minutes, and the same be first moistened on their inward sides with a clean cloth dipped into oil of oranges or spirit of turpentine, and a drop or two of the oil or spirit be let fall upon the lower glass at the other, so soon as the upper glass is laid down upon the lower, so as to touch it at one end as above and to touch the drop at the other end, making with the lower glass an angle of about 10 or 15 minutes, the drop will begin to move toward the concourse of the glasses and will continue to move with an accelerated motion till it arrives at that concourse of the glasses. For the two glasses attract the drop and make it run that way toward which the attractions incline. And if when the drop is in motion you lift up that end of the glasses where they meet and toward which the drop moves, the drop will ascend between the glasses, and therefore is attracted. And as you lift up the glasses more and more, the drop will ascend slower and slower, and at length rest, being then carried downward by its weight as much as upward by the attraction. And by this means you may know the force by which the drop is attracted at all distances from the concourse of the glasses.

Now by some experiments of this kind (made by Mr. Hauksbee) it has been found that the attraction is almost reciprocally in a duplicate proportion of the distance of the middle of the drop from the concourse of the glasses, *viz.*, reciprocally in a simple proportion by reason of the spreading of the drop and its touching each glass in a larger surface, and again reciprocally in a simple proportion by reason of the attractions growing stronger within the same quantity of attracting surface. The attraction therefore within the same quantity of attracting surface is reciprocally as the distance between the glasses. And therefore where the distance is exceeding small, the attraction must be exceeding great. By the table in the second part of the Second Book, where in the thicknesses of colored plates of water between two glasses are set down, the thickness of the plate where it appears very black is three eighths of the ten-hundred-thousandth part of an inch. And where the oil of oranges between the glasses is of this thickness, the attraction collected by the foregoing rule seems to be so strong, as within a circle of an inch in diameter, to suffice to hold up a weight equal to that of a cylinder of water of an inch in diameter and two or three furlongs in length. And where it is of a less thickness, the attraction may be proportionally greater and continue to increase until the thickness do not exceed that of a single particle of the oil. There are therefore agents in nature able to make the particles of bodies stick together by very strong attractions. And it is the business of experimental philosophy to find them out.

Now the smallest particles of matter may cohere by the strongest attractions and compose bigger particles of weaker virtue; and many of these may cohere and compose bigger particles whose virtue is still weaker, and so on for divers successions, until the progression end in the biggest particles on which the operations in chemistry and the colors of natural bodies depend, and which by cohering compose bodies of a sensible magnitude. If the body is compact and bends or yields inward to pression without any sliding of its parts, it is hard and elastic, returning to its figure with a force rising from the mutual attraction of its parts. If the parts slide upon one another, the body is malleable or soft. If they slip easily and are of a fit size to be agitated by heat, and the heat is big enough to keep them in agitation, the body is fluid, and if it be apt to stick to things it is humid; and the drops of every fluid affect a round figure by the mutual attraction of their parts, as the globe of the earth and sea affects a round figure by the mutual attraction of its parts by gravity.

Since metals dissolved in acids attract but a small quantity of the acid, their attractive force can reach but to a small distance from them. And as in algebra, where affirmative quantities vanish and cease there negative ones begin, so in mechanics, where attraction ceases there a repulsive virtue ought to succeed. And that there is such a virtue seems to follow from the reflections and inflections of the rays of light. For the rays are repelled by bodies in both these cases without the immediate contact of the reflecting or inflecting body. It seems also to follow from the emission of light, the ray, so soon as it is shaken off from a shining body by the vibrating motion of the parts of the body and gets beyond the reach of attraction, being driven away with exceeding great velocity. For that force which is sufficient to turn it back in reflection may be sufficient to emit it. It seems also to follow from the production of air and vapor, the particles when they are shaken off from bodies by heat or fermentation, so soon as they are beyond the reach of the attraction of the body, receding from it and also from one another with great strength, and keeping at a distance so as sometimes to take up above a million of times more space than they did before in the form of a dense body. Which vast contraction and expansion seems unintelligible by feigning the particles of air to be springy and ramous, or rolled up like hoops, or by any other means than a repulsive power. The particles of fluids which do not cohere too strongly and are of such a smallness as renders them most susceptible of those agitations which keep liquors in a fluor are most easily separated and rarified into vapor, and in the language of the chemists they are 'volatile,' rarifying with an easy heat and condensing with cold. But those which are grosser, and so less susceptible of agitation, or cohere by a stronger attraction are not separated without a stronger heat, or perhaps not without fermentation. And these last are the bodies which chemists call 'fixed,' and being rarified by fermentation become true permanent air, those particles receding from one another with the greatest force and being most difficultly brought together which upon contact cohere most strongly. And because the particles of permanent air are grosser and arise from denser substances than those of vapors, thence it is that true air is more ponderous than vapor, and that a moist atmosphere is lighter than a dry one, quantity for quantity. From the same repelling power it seems to be that flies walk upon the water without wetting their feet; and that the object glasses of long telescopes lie upon one another without touching; and that dry powders are difficultly made to touch one another so as to stick together, unless by melting them or wetting them with water, which by exhaling may bring them together; and that two polished marbles, which by immediate contact stick together, are difficultly brought so close together as to stick.

And thus nature will be very conformable to herself and very simple, performing all the great motions of the heavenly bodies by the attraction of gravity which intercedes those bodies and almost all the small ones of their particles by some other attractive and repelling powers which intercede the particles. The *vis inertiae* is a passive principle by which bodies persist in their motion or rest, receive motion in proportion to the force impressing it, and resist as much as they are resisted. By this principle alone there never could have been any motion in the world. Some other principle was necessary for putting bodies into motion; and now they are in motion, some other principle is necessary for conserving the motion. For from the various composition of two motions, it is very certain that there is not always the same quantity of motion in the world. For if two globes, joined by a slender rod, revolve about their common center of gravity with a uniform motion, while that center moves on uniformly in a right line drawn in the plane of their circular motion, the sum of the motions of the two globes, as often as the globes are in the right line described by their common center of gravity, will be bigger than the sum of their motions when they are in a line perpendicular to that right line. By this instance it appears that motion may be got or lost. But by reason of the tenacity of fluids and attrition of their parts, and the weakness of elasticity in solids, motion is much more apt to be lost than got and is always upon the decay. For bodies which are either absolutely hard or so soft as to be void of elasticity will not rebound from one another. Impenetrability makes them only stop. If two equal bodies meet directly *in vacuo*, they will by the laws of motion stop where they meet and lose all their motion, and remain in rest unless they be elastic and receive new motion from their spring. If they have so much elasticity as suffices to make them rebound with a quarter or half or three quarters of the force with which they come together, they will lose three quarters or half, or a quarter of their motion. And this may be tried by letting two equal pendulums fall against one another from equal heights. If the pendulums be of lead or soft clay, they will lose all or almost all their motions; if of elastic bodies, they will lose all but what they recover from their elasticity. If it be said that they can lose no motion but what they communicate to other bodies, the consequence is that *in vacuo* they can lose no motion, but when they meet they must go on and penetrate one another's dimensions. If three equal round vessels be filled, the one with water, the other with oil, the third with molten pitch, and the liquors be stirred about alike to give them a vortical motion, the pitch by its tenacity will lose its motion quickly, the oil being less tenacious will keep it longer, and the water being less tenacious will keep it longest but yet will lose it in a short time. Whence it is easy to understand that if many contiguous vortices of molten pitch were each of them as large as those which some suppose to revolve about the sun and fixed stars, yet these and all their parts would, by their tenacity and stiffness, communicate their motion to one another till they all rested among themselves. Vortices of oil or water, or some fluider matter, might continue longer in motion, but unless the matter were void of all tenacity and attrition of parts and communication of motion (which is not to be supposed) the motion would constantly decay. Seeing therefore the variety of motion which we find in the world is always decreasing, there is a necessity of conserving and recruiting it by active principles, such as are the cause of gravity, by which planets and comets keep their motions in their orbs and bodies acquire great motion in falling, and the cause of fermentation, by which the heart and blood of animals are kept in perpetual motion and heat, the inward parts of the earth are constantly warmed and in some places grow very hot, bodies burn and shine, mountains take fire, the caverns of the earth are blown up, and the sun continues violently hot and lucid and warms all things by his light. For we meet with very little motion in the world besides what is owing to these active principles. And if it were not for these principles the bodies of the earth, planets, comets, sun, and all things in them would grow cold and freeze, and become inactive masses; and all putrefaction, generation, vegetation, and life would cease, and the planets and comets would not remain in their orbs.

All these things being considered, it seems probable to me that God in the beginning formed matter in solid, massy, hard, impenetrable, movable particles, of such sizes and figures, and with such other properties and in such proportion to space as most conduced to the end for which he formed them; and that these primitive particles being solids are incomparably harder than any porous bodies compounded of them, even so very hard as never to wear or break in pieces, no ordinary power being able to divide what God himself made one in the first creation. While the particles continue entire, they may compose bodies of one and the same nature and texture in all ages; but should they wear away or break in pieces, the nature of things depending on them would be changed. Water and earth, composed of old worn particles and fragments of particles, would not be of the same nature and texture now, with water and earth composed of entire particles in the beginning. And therefore, that nature may be lasting, the changes of corporeal things are to be placed only in the various separations and new associations and motions of these permanent particles; compound bodies being apt to break, not in the midst of solid particles, but where those particles are laid together and only touch in a few points.

It seems to me further that these particles have not only a *vis inertiae*, accompanied with such passive laws of motion as naturally result from that force, but also that they are moved by certain active principles, such as is that of gravity and that which causes fermentation and the cohesion of bodies. These principles I consider, not as occult qualities supposed to result from the specific forms of things, but as general laws of nature by which the things themselves are formed, their truth appearing to us by phenomena, though their causes be not yet discovered. For these are manifest qualities, and their causes only are occult. And the Aristotelians gave the name of 'occult qualities,' not to manifest qualities, but to such qualities only as they supposed to lie hid in bodies and to be the unknown causes of manifest effects, such as would be the causes of gravity, and of magnetic and electric attractions, and of fermentations, if we should suppose that these forces or actions arose from qualities unknown to us and incapable of being discovered and made manifest. Such occult qualities put a stop to the improvement of natural philosophy, and therefore of late years have been rejected. To tell us that every species of things is endowed with an occult specific quality by which it acts and produces manifest effects is to tell us nothing, but to derive two or three general principles of motion from phenomena, and afterward to tell us how the properties and actions of all corporeal things follow from those manifest principles, would be a very great step in philosophy, though the causes of those principles were not yet discovered; and therefore I scruple not to propose the principles of motion above mentioned, they being of very general extent, and leave their causes to be found out.

Now by the help of these principles all material things seem to have been composed of the hard and solid particles above mentioned, variously associated in the first Creation by the counsel of an intelligent Agent. For it became him who created them to set them in order. And if he did so, it is unphilosophical to seek for any other origin of the world or to pretend that it might arise out of a chaos by the mere laws of nature, though being once formed it may continue by those laws for many ages. For while comets move in very eccentric orbs in all manner of positions, blind fate could never make all the planets move one and the same way in orbs concentric, some inconsiderable irregularities excepted which may have risen from the mutual actions of comets and planets upon one another, and which will be apt to increase till this system wants a reformation. Such a wonderful uniformity in the planetary system must be allowed the effect of choice. And so must the uniformity in the bodies of animals, they having generally a right and a left side shaped alike, and on either side of their bodies two legs behind and either two arms or two legs or two wings before upon their shoulders, and between their shoulders a neck running down into a backbone and a head upon it, and in the head two ears, two eyes, a nose, a mouth, and a tongue, alike situated. Also the first contrivance of those very artificial parts of animals, the eyes, ears, brain, muscles, heart, lungs, midriff, glands, larynx, hands, wings, swimming bladders, natural spectacles, and other organs of sense and motion, and the instinct of brutes and insects can be the effect of nothing else than the wisdom and skill of a powerful ever-living Agent, who being in all places is more able by his will to move the bodies within his boundless uniform sensorium, and thereby to form and reform the parts of the universe, than we are by our will to move the parts of our own bodies. And yet we are not to consider the world as the body of God, or the several parts thereof as the parts of God. He is a uniform being, void of organs, members, or parts, and they are his creatures subordinate to him, and subservient to his will; and he is no more the soul of them than the soul of man is the soul of the species of things carried through the organs of sense into the place of its sensation, where it perceives them by means of its immediate presence, without the intervention of any third thing. The organs of sense are not for enabling the soul to perceive the species of things in its sensorium, but only for conveying them thither; and God has no need of such organs, he being everywhere present to the things themselves. And since space is divisible *in infinitum* and matter is not necessarily in all places, it may be also allowed that God is able to create particles of matter of several sizes and figures, and in several proportions to space, and perhaps of different densities and forces, and thereby to vary the laws of nature and make worlds of several sorts in several parts of the universe. At least, I see nothing of contradiction in all this.

As in mathematics, so in natural philosophy, the investigation of difficult things by the method of analysis ought ever to precede the method of composition. This analysis consists in making experiments and observations, and in drawing general conclusions from them by induction, and admitting of no objections against the conclusions but such as are taken from experiment, or other certain truths. For hypotheses are not to be regarded in experimental philosophy. And although the arguing from experiments and observations by induction be no demonstration of general conclusions, yet it is the best way of arguing which the nature of things admits of, and may be looked upon as so much the stronger by how much the induction is more general. And if no exception occur from phenomena, the conclusion may be pronounced generally. But if at any time afterward any exception shall occur from experiments, it may then begin to be pronounced with such exceptions as occur. By this way of analysis we may proceed from compounds to ingredients and from motions to the forces producing them, and in general from effects to their causes and from particular causes to more general ones, till the argument end in the most general. This is the method of analysis; and the synthesis consists in assuming the causes discovered and established as principles, and by them explaining the phenomena proceeding from them and proving the explanations.

In the two first books of these *Optics*, I proceeded by this analysis to discover and prove the original differences of the rays of light in respect of refrangibility, reflexibility, and color, and their alternate fits of easy reflection and easy transmission, and the properties of bodies, both opaque and pellucid, on which their reflections and colors depend. And these discoveries being proved may be assumed in the method of composition for explaining the phenomena arising from them, an instance of which method I gave in the end of the First Book. In this Third Book I have only begun the analysis of what remains to be discovered about light and its effects upon the frame of nature, hinting several things about it and leaving the hints to be examined and improved by the further experiments and observations of such as are inquisitive. And if natural philosophy in all its parts, by pursuing this method, shall at length be perfected, the bounds of moral philosophy will be also enlarged. For so far as we can know by natural philosophy what is the first cause, what power he has over us, and what benefits we receive from him, so far our duty toward him, as well as that toward one another, will appear to us by the light of nature. And no doubt, if the worship of false gods had not blinded the heathen, their moral philosophy would have gone farther than to the four cardinal virtues; and instead of teaching the transmigration of souls, and to worship the sun and moon and dead heroes, they would have taught us to worship our true Author and Benefactor, as their ancestors did under the government of Noah and his sons before they corrupted themselves.

Notes

The quotation from Newton preceding the text of the selections is from a manuscript entitled, "A Scheme for Establishing the Royal Society." Quoted in Brewster, *Memoirs of Sir Isaac Newton,* Vol. I, p. 102.

In order to retain their seventeenth-century flavor, quotations in these Notes have been rendered in their original form, except for indiscriminate capitalization. Any other changes in the wording have been indicated by brackets.

[1] This letter to Oldenburg, then secretary of the Royal Society, was a reply by Newton to certain objections raised against his first scientific paper, *The New Theory about Light and Colors,* by Gaston Pardies. Like Robert Hooke, who also criticized this paper, Pardies called Newton's findings on the refrangibility of light a "hypothesis." Newton consequently, here and throughout his later writings, took pains to distinguish for his readers between experimental findings and results on the one hand, and speculative conjectures lacking experimental proof on the other. These latter he calls 'hypotheses.' It is a mistake to think that because his writings are full of scattered remarks to the effect that hypotheses are to be avoided whenever possible, or that because he does not rely upon hypotheses, Newton did not fully appreciate the experimental spirit or the role of hypothetical proposals in scientific inquiries. Newton cannot be accused of overemphasizing the function of mathematics and deduction in science on this point. For he makes it clear that 'hypotheses,' in his sense of the word, may aid in suggesting—they may "furnish" —experiments. Explanations, however, are to be based on experiments, not on conjectures; and when and if we can produce nothing more than experimentally unsupported suppositions, or 'hypotheses' in his sense, we have neither truth about the subject in question nor scientific knowledge. Thus, for example, Newton begins his *Optics* with the words:

My design in this book is not to explain the properties of light by hypotheses, but to propose and prove them by reason and experiments.

While he insists that hypotheses should not be preferred to experimental results, where such results may not be available or accessible to deciding a question, Newton does not altogether

181

condemn putting forth hypotheses. The fact is that Newton had a genius for advancing very bold and wonderfully speculative hypotheses himself when he was moved to do so. His letter to Boyle (Part IV) gives full evidence of this. Yet questionable as some of his speculations may have been, Newton rarely confused these with the rigorous discipline and consequences involved in establishing scientific explanations.

[2] These letters, written in March, 1713, as the second edition of the *Principia* was being prepared for the press, offer perhaps the best and clearest statements by Newton as to what he means by 'hypothesis.' The comparison of the function of hypotheses in geometry with the use of the same in experimental science is particularly instructive and typical of the mathematical outlook (and use of mathematics as a model to which science should conform) which to a large extent dominated seventeenth- and eighteenth-century science. Roger Cotes was at the time assisting in the more mundane tasks required in putting forth the second edition of the *Principia*. The various revisions and corrections were directed by Newton, but prepared for the printer by Cotes. The correspondence that thus ensued between the two men is given in Edleston's book. For this second edition of the *Principia*, Newton prepared his famous General Scholium (see Part III) on God and his place in natural philosophy. The latter part of this scholium contains Newton's well-known remarks on method and his frequently cited words, "I frame no hypotheses." Newton's directions to Cotes "to conclude the next paragraph in this manner" refer to the preparation of the General Scholium.

It is not in the rejection of hypotheses, once Newton's meaning of 'hypotheses' is understood, that later generations have called into question some of the underlying conceptions of Newton and his age as to the nature of scientific inquiry. Newton was deeply influenced by Galileo, Bacon, and Boyle concerning the importance of experiment, and was also equally impressed by the Cartesian outlook which regarded mathematics as the key and truly rational method by which nature was to be studied and known. Thus there is a strong strain of rationalism in Newton's writings which often appears, for example, in such statements as that the "first principles" of natural philosophy—such as the laws of motion—are "deduced from phenomena and afterward made general by induction." The assumption that an examination of phenomena will reveal indubitable principles or that the fundamental principles of science can be said to be *deduced* from phenomena has been seriously questioned in recent times. (It is worth noticing that in some cases Newton used the words 'derive' and 'infer,' instead of 'deduce,' in speaking of how these principles are obtained.) This spirit of scientific rationalism in the interpretation of the nature and origins, the beginnings and first principles, of science has been questioned in the case of Bacon, who thought that with a merely industrious collection of facts according to certain rules the causes and principles involved would become obvious. It has again been questioned in Descartes, who held that we must begin with intuitively (logically) "clear and distinct ideas," and that all of science should proceed in deductive fashion, like mathematics, to issue forth in a logically organized system of necessary truths. Somewhere between these two conceptions of the proper nature of scientific inquiry we may place Newton's statements that the first principles of science are deduced from phenomena and then made general by induction. For a valuable and penetrating historical study of the fundamental direction and problems in Newton's outlook, see J. H. Randall, Jr., "Newton's Natural Philosophy: Its Problems and Consequences," *Philosophical Essays in Honor of Edgar Arthur Singer, Jr.* See also E. A. Burtt, *The Metaphysical Foundations of Modern Physical Science*, pp. 202-23; E. W. Strong, "Newton's 'Mathematical Way,'" *Journal of the History of Ideas*, Vol. XII, No. I, 1951.

3 The scholium on absolute time, space, and motion has occasioned a great wealth of scientific and philosophic literature, representing a wide range of points of view and various interests. Some of these, along with further bibliographical material of interest, are given in Cajori's historical and explanatory notes, in his appendix to the *Principia*. Cf. especially pp. 637-46.

4 The General Scholium was written for the second edition of the *Principia* (1713). Cotes was finishing his task of readying this edition for the printer when Newton wrote him on March 2, 1712/13, that he was enclosing "the scholium which I promised to send you, to be added to the end of the book." During the revision of this second edition, the controversy between Leibniz and Newton as to the validity of their respective claims to the invention of the calculus reached such a point that it soon broke out into a long, complicated, and bitter war involving most of the leading Continental and English scientists of the time. As L. T. More writes of this controversy:

. . . it is undoubtedly the bitterest and most notorious dispute in the history of science; it engaged its principles and their sympathisers in recriminations for a quarter of a century; it will never lose its interest and never be satisfactorily settled. [*Isaac Newton—A Biography*, p. 565.]

All of the difficulties and issues at stake, plus the questionable behavior on both sides of this controversy, indeed make it hard to present accurately and impartially. More, however, has succeeded, as well as anyone to date, in doing this; see *ibid.*, especially, pp. 565-607. On the Continent the men of science were almost unanimous in their support of Leibniz. When it was learned that a new edition of the *Principia* was to be brought forth, a number of severe criticisms of Newton's philosophy began to circulate. These attacks were not directed to the generally unquestionable mathematical portions of the *Principia* and the development there of the principle of gravitation, but had to do rather with Newton's claim that his system required no hypotheses, and was thus to be preferred to the Cartesian hypothesis of vortices. Newton was accused of holding that

. . . attraction at a distance was to be an essential property of matter; if so, then they argued, he had introduced an occult quantity, and so his philosophy was as hypothetical as the Cartesian hypothesis of vortices in an occult medium. [More, *op. cit.*, p. 552.]

By 'occult' is meant *hidden,* in the sense of not being available to sensory experience or to experimental observation. For Newton's understanding of the meaning of 'occult qualities,' see the latter part of Quest. 31, in Questions from the *Optics,* p. 176.

A correspondent writes to Cotes, in 1711, that

I have nothing of news to send you: only the Germans and French have in a violent manner attacked the philosophy of Sir Is. Newton, and seem resolved to stand by Descartes. . . . [Edleston, *op. cit.*, p. 210.]

Thus it is that Newton began his scholium by considering the "many difficulties" involved in Descartes' hypothesis. Cotes' preface to the second edition of the *Principia* was concerned to deal with some of these charges against Newton (see Part IV, 4, and Note 13, page 198). That Newton did not mean to assert that gravity was an essential property of matter is clear from his second letter to Bentley (see Part III, 2).

Theological objections to Newton's philosophy had also been raised. Bishop Berkeley, in Part I of his *The Principles of Human Knowledge* (1710), Sections 110 to 117, criticized the notions of absolute time, space, and motion. He argued that to be conceived as meaningful ideas at all, *time, space,* and *motion* are relative in nature. Treated as absolute, these notions may lead to atheism. Hence, in speaking of his argument against pure and absolute space, Berkeley says that:

. . . the chief advantage arising from it is that we are freed from that dangerous dilemma, to which several who have employed their thoughts on that subject imagine themselves reduced, to wit, of thinking either that Real Space is God, or else that there is something beside God which is eternal, uncreated, infinite, invisible, immutable. Both which may justly be thought pernicious and absurd notions. It is certain that not a few divines, as well as philosophers of great note, have, from the difficulty they found in conceiving either limits or annihilation of space, concluded it must be divine. And some of late have set themselves particularly to show the incommunicable attributes of God agree with it. Which doctrine, how unworthy soever it may seem of Divine Nature, yet I do not see how we can get clear of it, so long as we adhere to the received opinions. [Section 117.]

This was directed at Newton and at the Cambridge Platonists, who deeply influenced Newton in theological questions associated with the nature of space. Leibniz also criticized Newton for regarding God as a great mechanic keeping the universal machine he had created from occasionally breaking down and for identifying God with the sensorium of infinite space as, it was held, the Cambridge Platonists did. An interesting sidelight, illustrating the extent to which this attack on Newton was felt in England, is offered by the following story in the *Biographica Britannica* article on Newton:

Leibniz renewed the charge of irreligion against Newton and attempted to disparage his philosophy because of pique over fluxions. This blackening method had its effect. It is supposed that Pope added two lines to the *Denial* as a censure of Newton's philosophy:

"Philosophy that lean'd on Heaven before,
Shrinks to her hidden cause, and is no more."

Dr. Warburton remarks that if Pope's excellent friend Dr. Arbuthnot had been consulted, "So unjust a reflection had never disgraced so noble a satire." On the hint of Warburton, Pope changed the lines with great pleasure into a compliment, as they now stand, on that divine genius:

"Shrinks to her second cause, and is no more."

[Quoted in More, *op. cit.*, p. 555.]

Against these theological charges, Newton's scholium was designed to explain the nature and place of God in the mechanical system. Profoundly influenced by Henry More in his theological beliefs, Newton speaks, in the *Optics*, of space as the *divine sensorium;* space is that in which the power and will of God directs and controls the physical world. Space is not to be identified with God, for Newton or for Henry More. Newton says in this scholium that God "governs all things, and knows all that are or can be done. He is not eternity or infinity, but eternal and infinite; He is not duration or space, but he endures and is present. He endures forever, and is everywhere present; and by existing always and everywhere, he constitutes duration and space." God constitutes duration and space since "by the same necessity [as that he exists] he exists *always* and *everywhere*." Space is the "omnipresence of God" and "bodies find no resistance from the omnipresence of God" Newton says. But God has other attributes, and Newton lists them, which make it obviously wrong to regard Newton as asserting that space *is* God. Henry More comes closer to such an identification. In his *Antidote against Atheism*, More writes:

For if after the removal of corporeal matter out of the world, there will be still space and distance in which this very matter, while it was there, was also conceived to lie, and this distant space cannot be conceived to be something, and yet not corporeal, because neither impenetrable nor tangible, it must of necessity be a substance incorporeal necessarily and eternally existent of itself; which the clearer *Idea of a Being absolutely perfect* will more fully and punctually inform us to be the *self-subsisting God*. [Appendix, 2nd ed., p. 338. See also, L. T. More, *op. cit.*, p. 553.]

But God is something else for More too, and space is but one manifestation of God. In his *Enchiridion Metaphysicum*, More speaks of infinite extension:

. . . which is commonly held to be mere space, is in truth a certain substance . . . a certain rather confused and vague representation of the divine essence or essential presence, in so far as it is distinguished from his life and activities. [Ch. VIII, Par. 14.]

More also writes:

That spiritual object, which we call space, is only a passing shadow, which represents for us, in the weak light of our intellect, the true and universal nature of the continuous divine presence, till we are able to perceive it directly with open eyes and at a nearer distance. [*Opera Omnia* I, pp. 171, 1,675-9. The two above quotations are to be found in Burtt and were translated by him, *op. cit.*, p. 141.]

For further discussion of Newton's conception of God and his religious outlook, see Burtt, *op. cit.*, pp. 280-99; L. T. More, *op. cit.*, pp. 552-4, 608-47.

Aside from important matters pertaining to his theological beliefs and his remarks on gravity, some of the statements in this scholium concerning God are illustrative of Newton's view of the relation of the knowing mind to the objects of knowledge. Thus he says of God:

We have ideas of his attributes, but what the real substance of anything is we know not.

It is worth noticing how close this wording and the view expressed is to that of Locke, speaking about substance. Locke had written twenty-three years earlier of substance as "something" we know not what:

The idea, then, to which we give the *general* name substance, being nothing but the supposed, but unknown, support of those qualities we find existing. . . . [Ch. XXIII, 2.]

Or of God, Locke, like Newton, says:

For though in His own essence, which we certainly do not know (not knowing the real essence of a pebble, or a fly, or of ourselves), God be simple and uncompounded; yet, I think, I may say we have no other idea of Him but a complex one of existence, knowledge, power, happiness. . . . [Ch. XXIII, 34. Bk. II, Locke's *Essay Concerning Human Understanding*, 1690.]

The general doctrine reflected here, of primary and secondary qualities had been advanced by Galileo and stated clearly by Descartes. In his Sixth Meditation, for example, Descartes set the stage for much of the subsequent philosophizing of British empiricism:

. . . although in approaching fire I feel heat, and in approaching it a little too near I even feel pain, there is at the same time no reason in this which could persuade me that there is in the fire something resembling this heat any more than there is in pain something resembling it; all that I have any reason to believe from this is, that there is *something in it, whatever it may be,* which excites in me these sensations of heat and pain. [*Meditations on First Philosophy,* VI, 1641. The italics are mine.]

5 When Robert Boyle died (1691), he left a will providing the sum of fifty pounds a year to found a lectureship for the general purpose of showing that science and scientific discoveries constitute the best evidence and provide the truest defense of a divine Providence and the Christian religion. The specific arrangement was:

To settle an annual salary for some divine or preaching minister, who shall be enjoined to perform the offices following: 1. To preach eight Sermons in the year, for proving the Christian religion against notorious infidels, *viz.*, Atheists, Deists, Pagans, Jews, and Mahometans; not descending to any controversies that are among Christians themselves. . . .

Richard Bentley was chosen as the first lecturer under the terms of this bequest. In his sermons, delivered in 1692, and titled *A Confutation of Atheism*, Bentley made use of Newton's *Principia*. The attack on atheism was directed to the views of Hobbes and Spinoza. A general title page was prefixed to the published lectures, which read:

The Folly and Unreasonableness of Atheism demonstrated from The Advantage and Pleasure of a Religious Life, The Faculties of Human Souls, The Structure of Animate Bodies, and The Origin and Frame of the World, etc.

Newton's work, it was felt, had established the existence of design in the universe, which in turn provided proof of a divine Providence. When Bentley prepared his seventh and eighth Boyle lectures for publication, he corresponded with Newton on questions concerning gravity and the nature of the universe, as these had bearing upon the religious themes with which he was concerned. Newton's answers, as given in these letters to Bentley, are of great value in revealing his thoughts on questions which, had he not been prompted in this way, he might never have set down on paper. Bentley, who later became a warm admirer of Newton, was made Master of Trinity College in 1700. It was Bentley who succeeded in getting Newton to work on a second edition of the *Principia* and who arranged for the talented young Roger Cotes to assist Newton in this task. These letters are important, not only because Newton discloses his thoughts on the place and role of God in the mechanical world, but also for the speculations on gravity that are presented. That Newton was misunderstood, even by his friends, on the relation of gravity to matter is clear in the warning:

You sometimes speak of gravity as essential and inherent to matter. Pray, do not ascribe that notion to me. For the cause of gravity is what I do not pretend to know. . . .

⁶ Thomas Burnet was the senior proctor at Cambridge in 1668, the year that Newton was admitted as Major Fellow and received his Master of Arts degree. Burnet was the author of a much admired work, *Theoria Telluris Sacra* ("The Sacred Theory of the Earth, containing an account of the original of the Earth, and of

all the general changes which it hath already undergone, or is to undergo, till the consummation of all things," 1680). Before the publication of his book, Burnet wrote Newton asking him for his opinion of the theory it contained. The long and interesting letter given here is Newton's second and perhaps final one of this correspondence. Speculation as to the origins and creation of the world was a popular topic in the seventeenth century among scientists, philosophers, and men of letters. Newton's views on the creation out of chaos and the formation of the world are here clearly stated. Of interest also is Newton's device for avoiding difficulties of a philosophic or scientific sort in the Biblical account of creation. Moses, he says, was not attempting a "philosophical" or scientific description of "realities," nor a false one, but "he described realities in a language artificially adopted to the sense of the vulgar," that is, the uneducated, unscientific layman. In speaking of Moses *accommodating* his words to the understanding of the vulgar, Newton employs logically the same argument that Milton uses in his *theory of accommodation*. We cannot know God as he really is, says Milton, but only as he, in sacred writings, "condescending to accommodate himself to our capacities, has shown that he desires we should conceive" of him. [*De Doctrina Christiana*, 1, 2.] Valuable as this letter is in revealing to us Newton's thoughts on these questions, it is well to keep in mind his own acknowledgment of its rather speculative character. "I have not," he concludes the letter, "set down anything I have well considered or will undertake to defend."

7 For an interesting paper discussing Newton's anti-Trinitarianism and containing the suggestion that in religious conviction Newton was "a Judaic monotheist of the school of Maimonides," see J. M. Keynes' "Newton, the Man," *The Royal Society Newton Tercentenary Celebrations*. Keynes also discusses the little known fact that, for an important part of his life, Newton was seriously addicted to alchemy. Also of interest is E. W. Strong's "Newton and God," *Journal of the History of Ideas*, April, 1952.

8 As early as 1664, when he was twenty-two years old, Newton was making experimental observations with a prism and was also concerned with the problem of improving the refracting telescope. It is still an undecided historical question as to which of these two lines of interest led him to the discovery of the nature of light. By 1688, we know from an account given in a letter, Newton had constructed a small reflecting telescope containing a concave metal mirror instead of an object glass. The letter, which is of importance, reads as follows:

Trinity College, Cambridge,
February 23, 1688/9

Sir,

I promised in a letter to Mr. Ent to give you an account of my success in a small attempt I had then in hand: and it is this. Being persuaded of a certain way whereby the practical part of optics might be promoted, I thought it best to proceed by degrees, and make a small perspective first, to try whether my conjectures would hold good or not. The instrument that I made is but six inches in length, it bears something more than an inch aperture, and a plano-convex eye-glass whose depth is 1/6th or 1/7th part of an inch; so that it magnifies about forty times in diameter, which is more than any six-feet tube can do, I believe, with distinctness. But, by reason of bad materials, and for want of good polish, it represents not things so distinct as a six-feet tube will do; yet I think it will discover as much as any three or four-feet tube, especially if the objects be luminous. I have seen with it Jupiter distinctly round and his satellites, and Venus horned. Thus, Sir, I have given you a short account of this small instrument, which, though in itself contemptible, may yet be looked upon as an epitome of what may be done according to this way, for I doubt not but in time a six-feet tube may be made after this method, which will perform as much as any sixty or hundred-feet tube made after the common way; whereas I am persuaded, that were a tube made after the common way of the purest glass, exquisitely polished, with the best figure that any geometrician (Des Cartes, etc.) hath or can design, (which I believe is all that men have hitherto attempted or wished for), yet such a tube would scarce perform as much more as an ordinary good tube of the same length. And this, however it may seem a paradoxical assertion, yet it is the necessary consequence of some experiments, which I have made concerning the nature of light. [Quoted in L. T. More, *op. cit.* p. 68.]

Word of this telescope came to the Royal Society, and Newton sent one, which was exhibited for the Society in 1671 or 1672. This invention caused a great deal of interest. Newton was proposed as a candidate for membership in the Society and elected as a Fellow at this time. The full implications and importance of the reflecting telescope had to wait, partly because of the technical difficulties involved in its construction. Refracting telescopes of 160 feet in length were used by Compani, Divini, and Huygens. Encouraged by the enthusiasm shown for his work, in a long series of letters to Oldenburg Newton discussed the plans and problems connected with the construction of larger reflecting telescopes. The warm interest with which his invention was received also led Newton to announce something more of greater importance and only vaguely implied in the last sentence of the letter above. This was his discovery of the dispersion of light and the nature of color. The significance of these studies and some of the excitement he must have felt in disclosing these secrets are conveyed by his words to Oldenburg, on January 18, 1672, when he asks to be informed as to

. . . what time the Society continue their weekly meetings; because, if they continue them for any time, I am purposing them, to be considered of and examined, an account of a philosophical discovery, which induced me to the making of the said telescope, and which I doubt not but will prove much more grateful than the communication of that instrument, being in my judgment the oddest, if not the most considerable detection which hath hitherto been made in the operations of nature. [*Opera Omnia*, IV, p. 274.]

These are the eager words of the young genius who had waited until what he thought was the favorable moment to announce his discoveries to the world and win public acclaim. Perhaps because of the disappointment over the general reception of his communication, Newton never again exhibited any strong desire to show the world what he had accomplished. At least he never again so boldly heralded his scientific work to the world. The odd and "considerable detection" he speaks of is the discovery of the heterogeneous character of white light. The *New Theory about Light and Colors* is the presentation of this "philosophical discovery"; it is also Newton's first serious scientific paper. The actual construction of such a paper, so familiar in the sciences now, had few if any precedents in Newton's day. Thus, in addition to the historical importance of its contents, this essay stands as a model of clarity and organization, a classic in the literature of scientific inquiries. In an unusually concise fashion, as compared with the style of such papers at the time, by selecting a few well-chosen and relevant experiments Newton offers proof of the fact that light consists of a quantity of rays possessing different refrangibilities. He also establishes the correspondence between refrangibility and color; the most refrangible rays are violet, the least refrangible red. From mixtures of the rays of primary colors new colors are obtained; white as a primitive color was obtained by a mixture of all the other colors of the spectrum. Newton also discusses some of the consequences of these findings in the study of telescopes and microscopes.

As admirable and plausible as this paper may appear to the modern reader, its revolutionary significance was not appreciated at the time, even by those most competent and capable of appreciating it. This can be seen by noting the reactions to it. It was called a "hypothesis" by Hooke; it was for the most part misunderstood and its value lost on Pardies, the distinguished professor of natural philosophy in the College of Claremont, Paris. The reaction of Huygens was no better. A long series of disappointing controversies and explanatory letters concerning the ideas in this paper mark the period between it and the *Hypothesis* of 1675-6 (See Note 9, below). Newton was at once extremely timid, proud, and jealous of his work and reputation; he loathed controversies and public exchanges. For an account of these controversies and their effect upon Newton, see L. T. More, *op. cit.*, pp. 82-121; L. Rosenfield "La théorie des couleurs de Newton et ses adversaires," *Isis*, 9 (1927), pp. 44-65.

9 In this long paper submitted to the Royal Society, Newton reviews some of his past work and offers a hypothesis as to the nature of light, distinguishing it from a vibrating ether. He also speaks of some of his more recent experiments on diffraction. The hypothesis is speculative, as Newton himself makes clear. Descartes had held that an ether existed which, as a dense and subtle medium, filled the interstices of air as well as the pores of transparent bodies. This ether consisted of the minutest globules spread out in a packed continuous series; a motion propagated along by a succession of these globules was the basis of light and color. Hooke, in part, took this view and explained light as the agitation or vibration of the ether moving along a straight line from a given body to the eye and there causing the sensation of light. Newton (in a letter to Oldenburg, December 21, 1675), speaking of the difference between his own position and that of Hooke, says:

I have nothing in common with him, but the supposition that ether is a medium susceptible of vibrations. Of which supposition I make a very different use: he supposing it light itself, which I suppose it is not. This is as great a difference, as between him and Des Cartes.

Newton begins his paper by considering Hooke's hypothesis. He then proceeds to advance a more "comprehensive" hypothesis than these others. If it is hypotheses or speculation that has seemed wanting in his earlier papers, he seems to say, he now will supply the deficiency. "I shall not," he warns, "assume either this or any other hypothesis"; but, he goes on to add, "I shall sometimes . . . speak of it as if I assumed it and propounded it to be believed." Nonetheless he was to some extent inclined to accept the hypothesis he puts forth. For it appears again in his letter to Boyle (see Part IV), and in a letter to Halley shortly before publication of the *Principia* Newton still seemed to consider this hypothesis seriously, saying:

I there suppose that the descending spirit acts upon bodies here on the superficies of the earth with a force proportional to the superficies of the parts; which cannot be, unless the diminution of its velocity in acting upon the first parts of any body it meets with, be recomposed by the increase of its density arising from that retardation. Whether this be true is not material. It suffices that 'twas the hypothesis. Now if this spirit descend from above with uniform velocity, its density, and consequently, its force, will be reciprocally proportional to the square of its distance from the centre. But if it descend with accelerated motion, its density will everywhere diminish as much as its velocity increases; and so its force (according to the hypothesis) will be the same as before, that is, still reciprocally as the square of its distance from the centre. [Quoted in W. W. Ball, *An Essay on Newton's Principia,* p. 166. See also, Burtt, *op. cit.,* p. 272.]

The letter to Oldenburg introducing the hypothesis is likewise of interest:

Sir,
    I have sent you the papers I mentioned, by John Stiles. Upon reviewing them I find some things so obscure as might have deserved a further explication by schemes [diagrams]; and some other things I guess will not be new to you, though almost all was new to me when I wrote them. But as they are, I hope you will accept of them, though not worth the ample thanks you sent. I remember in some discourse with Mr. Hooke, I happened to say that I thought light was reflected, not by the parts of glass, water, air, or other sensible bodies, but by the same confine or superficies of the ethereal medium which refracts it, the rays finding some difficulty to get through it in passing out of the denser into the rarer medium, and a greater difficulty in passing out of the rarer into the denser; and so being either refracted or reflected by that superficies, as the circumstances they happened to be in at their incidence make them able or unable to get through it. And for confirmation of this, I said further, that I thought the reflection of light, at its tending out of glass into air, would not be diminished or weakened by drawing away the air in an air-pump, as it ought to be if they were the parts of air that reflected; and added, that I had not tried this experiment, but thought he was not unacquainted with notions of this kind. To which he replied, that the notion was new, and he would the first opportunity try the experiment I propounded. But upon reviewing the papers I sent you, I found it there set down for trial, which makes me recollect that about the time I was writing these papers, I had occasionally observed in an air-pump here at Christ's College, that I could not perceive the reflection of the inside of the glass diminished in drawing out the air. This I thought to mention, lest my former forgetfulness, through my having long laid aside my thoughts on these things, should make me seem to have set down for certain what I never tried.
    Sir, I had formerly proposed never to write any hypothesis of light and colours, fearing it might be a means to engage me in vain disputes; but I hope a declared resolution to answer nothing that looks like a controversy, unless possibly at my own time upon some by-occasion, may defend me from that fear. And therefore, considering that such an hy-

pothesis would much illustrate the papers I promised to send you, and having a little time this last week to spare, I have not scrupled to describe one, so far as I could on a sudden recollect my thoughts about it; not concerning myself, whether it should be thought probable or improbable, so it do but render the paper I send you, and others sent formerly more intelligible. You may see by the scratching and interlining it was done in haste; and I have not had time to get it transcribed, which makes me say I reserve a liberty of adding to it, and desire that you would return these and the other papers when you have done with them. I doubt there is too much to be read at one time, but you will soon see how to order that. At the end of the hypothesis you will see a paragraph, to be inserted as is there directed. I should have added another or two, but I had not time, and such as it is I hope you will accept it.—Sir, I am your obedient servant,

Is. NEWTON

[Quoted in Brewster, *op. cit.*, Vol. I, pp. 132-4.]

[10] These selections from the *Optics* and the two letters following are interesting studies by Newton on the nature of perception. Generally, the "causal theory of perception" is put forth. Newton clearly conducts his analysis by accepting the doctrine of primary and secondary qualities. (On this point see Burtt, *op. cit.*, pp. 228-37.) Although Newton is here directly concerned with questions involving our perception of light and colors, the same analysis, in terms of outer objects as causes of sensations within us, or within our "sensorium," holds also for sound and the other senses.

[11] Everyone knows the story of Newton's being struck by an apple while sitting in his orchard in the summer of 1665/6 and responding by inventing the law of gravitation. It is probably nothing but a story, but the history of the *Principia* does begin at this time. For Newton was then working on the question as to whether or not the laws of planetary motion could be accounted for by calculating the force of attraction as varying inversely to the square of the distance between two bodies. Why Newton waited for twenty years before announcing his law is yet an open and fascinating question. The traditional answer, that in 1666 Newton employed an inaccurate value for the diameter of the earth, and that consequently his computations for what would be the intensity of gravity on the surface of the earth did not agree with his experimental results, is hardly to be accepted. See Cajori, "Newton's Twenty Years' Delay in Announcing the Law of Gravitation," *Sir Isaac Newton*, memorial volume published by the History of Science Society, Baltimore, Maryland, 1928; and Cajori's notes in the *Principia*, pp. 663-4. Cajori believes that Newton was delayed by

. . . theoretical questions of great difficulty relating to the attraction of a sphere upon an external point—a problem which he did not solve before 1684 or 1685, and which was first explained in the *Principia*, Book I, Proposition LXXV and LXXVI, and the Corollaries.

L. T. More, commenting on Cajori's thesis, states that the delay on Newton's part was primarily due to "temperamental procrastination and distaste for developing systematically any mathematical problem." [*Op. cit.*, p. 242.]

Briefly, the story as we know it is that, in 1666, Newton had found the force governing the motions of planetary bodies accounted for by Kepler's rule of periodic times. Newton did not announce this discovery. Perhaps this was because he was also engaged in other studies. Or, he might have waited until he could extend this initial finding to cases where the dimensions of bodies and the relation of the size of any bodies to their attraction could be determined. For Newton had been considering only the earth and moon, and the greatness of their distance allowed him to neglect their dimensions. At any rate, in 1673, Huygens formulated his laws of centrifugal force, and Newton realized that an important deduction from his still unannounced law of gravity had been anticipated. Still he did nothing about it. In 1674 Hooke wrote asking Newton to communicate anything of philosophic interest to the Society. Newton's reply was hardly one to please his former antagonist. To soften the effect of this letter, however, Newton later asked Halley to communicate a suggestion to the Society as to how the earth's rotation might be demonstrated. His idea was that, if a body were dropped from a considerable height, it would fall in a straight line if the earth did not rotate; but if the earth did rotate, the line of fall would not be quite straight. The Society was much interested in this notion, but Newton made an error in supposing that the path of the falling body would be a spiral. Nothing pleased Hooke more than to be able to point out this mistake publicly. Newton was forced to admit that he had been wrong, but this concession had the subsequent value of stimulating him to work again on the whole question. Newton apparently made great progress and further important discoveries, and then once more dropped the entire matter. Finally, in 1683, Halley, studying Kepler's third law, reached the decision that the centripetal force of attraction was inversely proportional to the square of the distance. He could not prove this, however, and discussed it in a meeting with Sir Christopher Wren and Hooke. Hooke claimed that this was the principle upon which all the laws of celestial motion were to be demonstrated. He claimed further that he could prove this. Wren put up a prize for the man who could give a proof of a planet obeying this law. Hooke said he would wait with his proof until other men had tried to solve the problem and failed; they would then appreciate what he had done. In some accounts, Hooke is reputed to have said that the path of the planet obeying the law would be an ellipse. If Hooke had this proof at the time (which is very doubtful), he made the greatest mistake in his life in holding it back. Halley proceeded to struggle with the problem for a time and then went to visit Newton at Cambridge. His first question, we are told, was to ask what would be the curve described by a planet, on the supposition that the gravity diminished as the square of the distance, to which Newton at once answered that it would be an ellipse. Halley was the more amazed when Newton stated that he had already calculated this, but was unable to find the papers to prove it, and promised to send them to Halley as soon as he could. Newton then started in again on the problem; the result was a series of written lectures called *De Motu Corporum*, which he gave in Cambridge in 1684 and 1685. Halley gave a report on the *De Motu* to the Royal Society in 1684. Much curiosity and interest was expressed; the Society asked Newton to send them a published version of his treatise when he could. The final response came from Newton: in April, 1686, he presented to the Society the manuscript of the *Principia*.

The history of science can produce nothing comparable to this case of a youth, in his twenties, who in two years made discoveries which were to rock his age and usher in a new era in science. In a memorandum (in the Portsmouth collection) written about 1714, Newton gives us an account—with some minor mistakes as to the exact dates—of these early accomplishments.

In the beginning of the year 1665 I found the method of approximating series and the rule for reducing any dignity of any binomial into such a series. In the same year, in May, I found the method of tangents . . . and in November had the direct method of fluxions, and the next year in January had the theory of colours and in May following I had entrance into the inverse method of fluxions. And the same year I began to think of gravity extending to the orb of the moon, and having found out how to estimate the force with which a globe revolving within a sphere presses the surface of the sphere, from Kepler's rule of the periodical times of the planets being in a sesquialterate proportion of their distances from the centres of their orbs I deduced that the forces which keep the planets in their orbs must [be] reciprocally as the squares of their distances from the centres about which they revolve: and thereby compared the force requisite to keep the moon in her orb with the force of gravity at the surface of the earth, and found them answer pretty nearly. All this was in the two plague years of 1665 and 1666, for in those days I was in the prime of my age for invention, and minded mathematics and philosophy more than at any time since. What Mr. Huygens has published since about centrifugal forces I suppose he had before me. At length in the winter between the years 1676 and 1677, I found the proposition that by a centrifugal force reciprocally as the square of the distance a planet must revolve in an ellipsis about the centre of the force placed in the lower umbilicus of the ellipsis and with a radius drawn to that centre describe areas proportional to the times. And in the winter between the years 1683 and 1684 this proposition with the demonstration was entered in the Register book of the R. Society.

The general formulation of the law of gravitation is as follows:

Every particle of matter attracts every other particle with a force varying directly as the product of their masses and inversely as the square of the distance between them.

This is not to be found, stated as such, in the *Principia* or in the *System of the World*. The selections given in the text are the nearest approach to the modern statement of the law; see also the words in the second to the last paragraph of the General Scholium (Part III, p. 45).

12 The famous letter of Newton to Robert Boyle on the supposition of a contracting and dilating ether and the cause of gravity is a further extension of the matters discussed in Newton's *Hypothesis* on light and colors (Part IV, 2). The letter is of interest not only because of the ideas it sets forth, but also because it allows us to follow Newton's train of thought on the basis of what he frankly acknowledges are "suppositions." Were it not for a feeling of indebtedness on his part, for which we can be grateful, it is likely that Newton would never have allowed himself to indulge in such very speculative hypotheses as those he here addresses to Boyle. But perhaps it was only to avoid any possibility of falling into a controversy and having to defend notions not themselves accessible to experimental proof that Newton takes pains to stress his dissatisfaction and reluctance in developing these ideas. The seductive temptation to speculate, to advance "hypotheses" beyond the range of what actual experimentation at any one time might warrant, is only normal; it is a sign of intelligence in life and a condition of progress in science. Newton was ever watchful, however, not to let "suppositions" be taken for or to interfere with experimentation. As late as 1717, Newton still insisted ". . . I do not take gravity for an essential property of bodies." The cause of gravity remained to be proposed as "a question, because I am not yet satisfied about it for want of experiments." [*Optics*, Advertisement II, 2nd ed., 1717.] Speculative

as it is, this letter leaves little doubt that Newton was seriously interested in entertaining and examining the particular ideas here expressed, and this in spite of the fact that "in natural philosophy . . . there is no end of fancying."

[13] Newton's General Scholium did not entirely satisfy Bentley and Cotes. It did not, as L. T. More points out,

. . . sufficiently crush the Cartesians or manifest the glory of the *Principia*; it softened the denial that occult qualities had been introduced; it was silent as to Leibniz and the invention of the calculus. Most important of all it did not hurl back with scorn the charge of the materialism of his philosophy, and the irreligion of its author, which Leibniz had insinuated in the ear of that royal blue stocking, the Princess of Wales, who had just come to England from Hanover and the teaching of Leibniz. [*Op. cit.*, p. 555.]

Bentley conferred first with Cotes and then with Newton concerning these seemingly urgent and important matters. It was decided that Cotes should write a preface in which he could deal with these questions. In a joint letter from Newton and Bentley to Cotes, Bentley writes:

I have Sir Isaac's leave to remind you of what you and I were talking of, an alphabetical index, and a preface in your own name; if you please to draw them up ready for the press, to be printed after my return to Cambridge, you will oblige. Yours. R. Bentley.

Cotes, in turn, agreed to do the index and wrote to Bentley asking what Newton wanted in the preface. The Leibniz-Newton controversy over the invention of the calculus was in full swing; Cotes suggested dealing with this, presenting a full defense of Newton along the lines already put forth by the Royal Society in the *Commercium Epistolicum*. He further suggested that Newton, or Newton and Bentley together, should write the preface and, knowing Newton's reluctance to engage personally in public controversies, added: "You may depend upon it that I will own it and defend it as well as I can if hereafter there should be occasion." There was no getting out of it this way, however, and Bentley let Cotes know this much in a characteristic reply:

At Sir Isaac Newton's March 12.
Dear Sir,

I communicated your letter to Sir Isaac, who happened to make me a visit this morning, and we appointed to meet this evening at his house, and there to write you an answer. For the close of your letter, which proposes a preface to be drawn up here, and to be fathered by you, we will impute it to your modesty; but you must not press it further, but go about it yourself. For the subject of the preface, you know it must be to give an account, first of the work itself, 2dly of the improvements of the new edition; and then you have Sir Isaac's consent to add what you think proper about the controversy of the first invention. You yourself are full master of it, and want no hints to be given you. However when it is drawn up, you shall have his and my judgement, to suggest any thing that may improve it. Tis both our opinions, to spare the name of M. Leibniz, and abstain from all words or epithets of reproach; for else, that will be the reply (not that its untrue) but that its rude and uncivil. Sir Isaac presents his service to you.

I am Yours,
R. Bentley

[Quoted in Edleston, *op. cit.*, p. 150.]

Cotes then sent to Newton an outline of his plan for the preface in a letter of March 18, 1713.

Sr.,
I have Dr. Bentley's letter. . . . I think it will be proper besides the account of the book and its improvements to add something more particularly concerning the manner of philosophising made use of and wherein it differs from that of Descartes and others, I mean first demonstrating the principle it employs. This I would not only assert but make evident by a short deduction of the principle of gravity from the phenomena of nature in a popular way that it may be understood by ordinary readers. . . .
After this specimen I think it will be proper to add some things by which your book may be cleared from some prejudices which have been industriously laid against it. As that it deserts mechanical causes, is built upon miracles, and recurrs to occult qualities. . . . I do not propose to mention Mr. Leibnitz's name; twere better to neglect him, but the objections I think may very well be answered and even retorted upon the maintainers of vortices. . . . [Quoted in *ibid.*, pp. 151-4.]

In a brief reply, Newton described some matters he wished to have printed beneath his original *Praefatio ad Lectorem* and, in a postscript of some historical importance, adds: "If you write any further preface, I must not see it, for I find I shall be examined about it." Thus were Cotes' first plans for the preface considerably modified and the spirited initial ideas made rather tame. The controversy over the invention of the calculus is not referred to; Leibniz's name is not mentioned, but his objections to Newton and the Cartesian theory of vortices are discussed. Cotes finished his task alone, and Newton, we can assume, never saw more of the preface until after the second edition had come out. Enough was said in Cotes' preface, however, to have aroused Leibniz to call it *"pleine d'aigreur"* (in a letter of April 9, 1716).

Of some historical interest concerning Cotes' preface is the fact that it has frequently been interpreted as advocating "action at a distance" and that gravity is an essential property of matter. As to the first of these points, Cotes never uses the words "action at a distance"; nor does he assert that celestial space is "void of all matter," as Samuel Clarke did. (For some significant remarks by Newton on action at a distance, see 29 and 31 in Questions from the *Optics*.) But he does say, and this is the passage which has been taken as advocating action at a distance:

Those who would have the heavens filled with a fluid matter, but suppose it void of any inertia, do indeed in words deny a vacuum, but allow it in fact. For since a matter of that kind can noways be distinguished from empty space, the dispute is now about the names and not the natures of things . . . .

(See Cajori's notes, *op. cit.*, for a further discussion of this point in Cotes' preface). On the second point, Cotes does say:

. . . gravity is found in all bodies universally [and] . . . either gravity must have a place among the primary qualities of all bodies, or extension, mobility, and impenetrability must not.

Cotes may thus have unwittingly contributed to the misunderstanding of Newton's views on this matter. Somewhat later, in the *Optics* (1717), Newton stated his position with his usual caution:

. . . to show that I do not take gravity for an essential property of bodies, I have added one Question concerning its cause, choosing to propose it by way of a Question because I am not yet satisfied about it for want of experiments. [*Optics*, Advertisement II, 2nd ed., 1717.]

We know that Newton had to take pains to assure even such followers as Bentley that he did not regard gravity as "essential and inherent to matter." "Pray do not ascribe that notion to me," Newton writes (in his letters to Bentley, Part III, 2). Yet there are statements in the *Principia,* too, which could easily be taken to imply "that notion." If Newton had read Cotes' finished preface, the misleading phrases might have been changed. Nonetheless, although Cotes' preface did encourage this misinterpretation of Newton, we know for a fact that Cotes did not, as is sometimes claimed, advocate this view himself. For he showed the preface, before publication, to Dr. Samuel Clarke. Clarke drew attention to this very point and, in a return letter on June 25, 1713, Cotes writes:

Sr.,
I received your very kind letter. I return you my thanks for your corrections of the Preface, and particularly for advice in relation to that place where I seemed to assert gravity to be essential to bodies. . . . I therefore struck it out immediately . . . my design in that passage was not to assert gravity to be essential to matter, but rather to assert that we are ignorant of the essential properties of matter and that in respect of our knowledge gravity might lay as fair a claim to that title as the other properties which I mentioned. . . . [Cambridge, June 25, 1713. Quoted in Edleston, *op. cit.*, pp. 158-9.]

[14] In Advertisement II to the second edition of the *Optics* (July 16, 1717), Newton wrote:

. . . at the end of the Third Book I have added some Questions. And to show that I do not take gravity for an essential property of bodies, I have added one Question concerning its cause, choosing to propose it by way of a Question because I am not yet satisfied about it for want of experiments.

In this series of brilliantly suggestive "Queries," Newton lets his thought range over a group of topics: the nature of light, gravity and matter, the proper method of natural inquiry, the evidence for and knowable aspects of God. In an extremely condensed fashion, Newton appears here almost to be summarizing, in the actual order of their development, the fundamental problems and ideas which had occupied him most of his life. By a careful study of the full significance of the thoughts motivating these Questions and reflected in them, we may come as close perhaps as creatures of another age possibly can to an intimate sense of the workings of that unique, strange, and difficult mind of unsurpassed genius.

[15] The *theory of fits* is advanced in order to provide an explanation of the colors of thin plates and to account for the fact that rays of light "are by some cause or other alternately disposed to be reflected or refracted for many vicissitudes." The theory, as put forth in Query 17 and further developed in Queries 28 and 29, is stated in such a way as to be compatible with a wave or undulatory version of the nature of light. Newton's *interval of fits* between two successive *fits of easy transmission* (see his definition below) would vary with the color of the light in question; the interval is greatest for red light and smallest for violet. Thus the length of these intervals of fits is in some respects akin to what, in the wave theory of light, would be termed the 'wave length' of the light. Newton's own "composite-wave" hypothesis as to the nature of light retains elements of both a wave and particle theory (see L. T. More, *op. cit.*, pp. 112-21). It would not be correct to say that Newton definitely held to only one

of these two possible conceptions of light. Experimental evidence did not permit a decided choice to be made, and Newton was as cautious as the scientists today who still face the same choice. (There is a valuable discussion by I. Bernard Cohen on the question of the *Optics* and these two theories of light, as well as on the larger theme of the significance of the *Optics* as scientific literature. See his preface to the recent republication of the *Opticks,* Dover Publications, New York 1952, pp. ix-lviii.) Newton's theory of Fits, it must be observed, is an alternative to Hooke's theory of an interference of ethereal waves. In Hooke's *Micrographia* (1665), a wave theory of light is proposed. Although Hooke's name is not mentioned, Part III of the *Optics* shows Newton to be heavily indebted to the book Hooke had written nearly fifty years earlier.

The theory of fits is best stated by Newton in Proposition XII, the definition following this, and Proposition XIII, Book II, Part III of the *Optics,* as follows:

PROPOSITION XII

Every ray of light in its passage through any refracting surface is put into a certain transient constitution or state, which in the progress of the ray returns at equal intervals, and disposes the ray at every return to be easily transmitted through the next refracting surface, and between the returns to be easily reflected by it.

. . . What kind of action or disposition this is, whether it consists in a circulating or a vibrating motion of the ray or of the medium or something else, I do not here inquire. Those that are averse from assenting to any new discoveries but such as they can explain by an hypothesis may for the present suppose that, as stones by falling upon water put the water into an undulating motion and all bodies by percussion excite vibrations in the air, so the rays of light, by impinging on any refracting or reflecting surface, excite vibrations in the refracting or reflecting medium or substance, and by exciting them agitate the solid parts of the refracting or reflecting body, and by agitating them cause the body to grow warm or hot; that the vibrations thus excited are propagated in the refracting or reflecting medium or substance, much after the manner that vibrations are propagated in the air for causing sound and move faster than the rays so as to overtake them; and that when any ray is in that part of the vibration which conspires with its motion it easily breaks through a refracting surface, but when it is in the contrary part of the vibration which impedes its motion it is easily reflected; and, by consequence, that every ray is successively disposed to be easily reflected or easily transmitted by every vibration which overtakes it. But whether this hypothesis be true or false I do not here consider. I content myself with the bare discovery, that the rays of light are by some cause or other alternately disposed to be reflected or refracted for many vicissitudes.

DEFINITION

The returns of the disposition of any ray to be reflected I will call its 'fits of easy reflection,' and those of its disposition to be transmitted its 'fits of easy transmission,' and the space it passes between every return and the next return the 'interval of its fits.'

PROPOSITION XIII

The reason why the surfaces of all thick transparent bodies reflect part of the light incident on them and refract the rest is that some rays at their incidence are in fits of easy reflection and others in fits of easy transmission.

[For brief accounts of this theory, see L. T. More, *op. cit.*, p. 115, and E. T. Whittaker's introduction to Newton's *Optics*, New York, 1931.]

Selected Bibliography

I. Sources

Isaaci Newtoni Opera quae exstant Omnia. Commentariis illustrabat Samuel Horsley. 5 vols. London, 1779-85.

Isaac Newton, *The Mathematical Principles of Natural Philosophy and His System of the World.* Motte's translation revised and supplied with an historical and explanatory appendix, by Florian Cajori. Berkeley, 1934.

*Isaac Newton, *Opticks: or, a Treatise of the Reflections, Refractions, Inflections and Colours of Light.* 4th ed., corrected, London, 1730.

Isaac Newton, "New Theory about Light and Colours," *Philosophical Transactions of the Royal Society.* No. 80, pp. 3075-87. Feb., 1672.

Brewster, Sir David, *Memoirs of the Life, Writings, and Discoveries of Sir Isaac Newton.* 2 vols. Edinburgh, 1850.

Edleston, J., *Correspondence of Sir Isaac Newton and Professor Cotes,* including letters of other eminent men. London, 1850.

II. Additional Sources and Works Cited in the Text and Notes

Ball, W. W. R., *An Essay on Newton's Principia.* London, 1893.

Bentley, R., *Eight Sermons against Atheism.* London, 1693.

Berkeley, G., *A Treatise Concerning the Principles of Human Knowledge.* Dublin, 1710.

Boyle, R., *The Works of the Honourable Robert Boyle,* edited by Thomas Birch. 6 vols. London, 1772.

Burtt, E. A., *The Metaphysical Foundations of Modern Physical Science.* 2nd ed., revised. London, 1932.

Cajori, F., "Newton's Twenty Years' Delay in Announcing the Law of Gravitation," *Sir Isaac Newton 1727-1927. A Bicentenary Evaluation of His Work.* The History of Science Society. Baltimore, 1928.

Cohen, I. B., Preface to Newton's *Opticks.* New York, 1952.

Descartes, R., *Meditationes de Prima Philosophia.* Paris, 1641.

Keynes, J. M., "Newton, the Man," *The Royal Society Newton Tercentenary Celebrations.* Cambridge, 1947.

*Locke, J., *An Essay Concerning Human Understanding.* London, 1690.

Milton, J., *De Doctrina Christiana*. (M.S., 1661.) London, 1825.

More, L. T., *Isaac Newton—A Biography*. New York, 1934.

Randall, J. H. Jr., "Newton's Natural Philosophy: Its Problems and Consequences," *Philosophical Essays in Honor of Edgar Arthur Singer, Jr.*, edited by Clarke and Nahm. Philadelphia, 1942.

Rosenfield, L., "La Théorie des Couleurs de Newton et ses Adversaires," *Isis*, 9, 1927.

Strong, E. W., "Newton's 'Mathematical Way,' " *Journal of the History of Ideas*. January, 1951.

————, "Newton and God," *Journal of the History of Ideas*. April, 1952.

Whittaker, E. T., Introduction to Newton's *Opticks*. New York, 1934.

III. Further Readings on Various Aspects of Newton's Thought

In view of the enormous amount of secondary literature available on Newton, the following bibliography is in no way to be considered complete. The attempt has been made merely to offer a list of helpful works for those interested in pursuing the study of Newton. Further references to the literature in question may be found in many of the works cited here.

A Catalogue of the Portsmouth Collection of Books and Papers written by or belonging to Sir Isaac Newton. Cambridge, 1888.

Ball, W. W. R., *History of the Study of Mathematics at Cambridge*. Cambridge, 1889.

Bloch, L., *La Philosophie de Newton*. Paris, 1908.

Broad, C. D., *Scientific Thought*, Chs. I-V. London, 1923.

*————, "Sir Isaac Newton," *Proceedings of the British Academy*. 1927.

Brodetsky, S., *Sir Isaac Newton*. London, 1927.

Brunschvicg, L., *L'Experience Humaine*, Ch. XXIV. Paris, 1922.

Cassirer, E., *Das Erkenntnis Problem*, Vol. II, Bk. 7. Berlin, 1911.

Dannemann, F., *Die Naturwissenschaften in ihrer Entwickelung*, Vol. II, Ch. XII. Leipzig, 1921.

De Morgan, A., *Essays on the Life and Work of Newton*. London, 1914.

Dewey, J., *The Quest for Certainty*, pp. 114-22, 142-4. New York, 1929.

————, *Reconstruction in Philosophy*, Chs. III, IV. New York, 1950.

Dingler, H., *Das Experiment*, Part III. Munich, 1928.

Duhem, P., *L'Evolution da la mécanique*. Paris, 1905.

Frost, W., *Bacon und die Natur-philosophie*, Part II, Chs. VI, VII. Munich, 1927.

*Hertz, H., *The Principles of Mechanics*, Introduction by H. Helmholtz. London, 1899.

History of Science Society, *Sir Isaac Newton 1727-1927*. Baltimore, 1928. (A group of studies on aspects of Newton's work. See the essays by G. D. Birkhoff, F. Cajori.)

Hoffding, H., *A History of Modern Philosophy*, Vol. I, Bk. IV, Ch. 3. London, 1900.

Mach, E., *The Science of Mechanics*. Chicago, 1919.

Maclaurin, C., *An Account of Sir Isaac Newton's Philosophical Discoveries*. London, 1748.

Mises, R. von, *Positivism*. Harvard, 1951.

Nicolson, M. H., *Newton Demands the Muse*. Princeton, 1946. (An interesting study of the effect of the *Optics* on eighteenth-century poetry.)

Pemberton, H., *A View of Sir Isaac Newton's Philosophy*. London, 1728.

Randall, J. H. Jr., *The Making of the Modern Mind*, Chs. XI, XII. New York, 1940.

Raphson, J., *History of Fluxions*. London, 1715.

Rigaud, S. P., *Correspondence of Scientific Men of the Seventeenth Century*. Oxford, 1841.

————, *Historical Essay on the First Publication of Sir I. Newton's Principia*. London, 1838.

Roberts, M., and Thomas, E. R., *Newton and the Origin of Colours*. London, 1934.

Rosenberger, F., *Newton und seine physikalischen Prinzipien*. Leipzig, 1895.

Royal Society Newton Tercentenary Celebrations. Cambridge, 1947. (Includes interesting essays by E. N. da C. Andrade, J. M. Keynes, N. Bohr, *et al.*)

Snow, A. J., *Matter and Gravity in Newton's Philosophy*. Oxford, 1926.

Whitehead, A. N., *Science and The Modern World*, Ch. III. New York, 1925.

————, "The First Physical Synthesis," *Science and Civilization*, edited by F. S. Marvin. London, 1928. (Reprinted in *Essays in Science and Philosophy*. New York, 1947.)

Whittaker, E. T., *A History of the Theories of Aether and Electricity, from the Age of Descartes to the Close of the Nineteenth Century*. London, 1910.

————, *From Euclid to Eddington*. Cambridge, 1949.

A CATALOG OF SELECTED
DOVER BOOKS
IN ALL FIELDS OF INTEREST

A CATALOG OF SELECTED DOVER
BOOKS IN ALL FIELDS OF INTEREST

CONCERNING THE SPIRITUAL IN ART, Wassily Kandinsky. Pioneering work by father of abstract art. Thoughts on color theory, nature of art. Analysis of earlier masters. 12 illustrations. 80pp. of text. 5⅜ x 8½. 23411-8

ANIMALS: 1,419 Copyright-Free Illustrations of Mammals, Birds, Fish, Insects, etc., Jim Harter (ed.). Clear wood engravings present, in extremely lifelike poses, over 1,000 species of animals. One of the most extensive pictorial sourcebooks of its kind. Captions. Index. 284pp. 9 x 12. 23766-4

CELTIC ART: The Methods of Construction, George Bain. Simple geometric techniques for making Celtic interlacements, spirals, Kells-type initials, animals, humans, etc. Over 500 illustrations. 160pp. 9 x 12. (Available in U.S. only.) 22923-8

AN ATLAS OF ANATOMY FOR ARTISTS, Fritz Schider. Most thorough reference work on art anatomy in the world. Hundreds of illustrations, including selections from works by Vesalius, Leonardo, Goya, Ingres, Michelangelo, others. 593 illustrations. 192pp. 7⅛ x 10¼. 20241-0

CELTIC HAND STROKE-BY-STROKE (Irish Half-Uncial from "The Book of Kells"): An Arthur Baker Calligraphy Manual, Arthur Baker. Complete guide to creating each letter of the alphabet in distinctive Celtic manner. Covers hand position, strokes, pens, inks, paper, more. Illustrated. 48pp. 8¼ x 11. 24336-2

EASY ORIGAMI, John Montroll. Charming collection of 32 projects (hat, cup, pelican, piano, swan, many more) specially designed for the novice origami hobbyist. Clearly illustrated easy-to-follow instructions insure that even beginning papercrafters will achieve successful results. 48pp. 8¼ x 11. 27298-2

THE COMPLETE BOOK OF BIRDHOUSE CONSTRUCTION FOR WOODWORKERS, Scott D. Campbell. Detailed instructions, illustrations, tables. Also data on bird habitat and instinct patterns. Bibliography. 3 tables. 63 illustrations in 15 figures. 48pp. 5¼ x 8½. 24407-5

BLOOMINGDALE'S ILLUSTRATED 1886 CATALOG: Fashions, Dry Goods and Housewares, Bloomingdale Brothers. Famed merchants' extremely rare catalog depicting about 1,700 products: clothing, housewares, firearms, dry goods, jewelry, more. Invaluable for dating, identifying vintage items. Also, copyright-free graphics for artists, designers. Co-published with Henry Ford Museum & Greenfield Village. 160pp. 8¼ x 11. 25780-0

HISTORIC COSTUME IN PICTURES, Braun & Schneider. Over 1,450 costumed figures in clearly detailed engravings–from dawn of civilization to end of 19th century. Captions. Many folk costumes. 256pp. 8⅜ x 11¾. 23150-X

CATALOG OF DOVER BOOKS

STICKLEY CRAFTSMAN FURNITURE CATALOGS, Gustav Stickley and L. & J. G. Stickley. Beautiful, functional furniture in two authentic catalogs from 1910. 594 illustrations, including 277 photos, show settles, rockers, armchairs, reclining chairs, bookcases, desks, tables. 183pp. 6½ x 9¼. 23838-5

AMERICAN LOCOMOTIVES IN HISTORIC PHOTOGRAPHS: 1858 to 1949, Ron Ziel (ed.). A rare collection of 126 meticulously detailed official photographs, called "builder portraits," of American locomotives that majestically chronicle the rise of steam locomotive power in America. Introduction. Detailed captions. xi+ 129pp. 9 x 12. 27393-8

AMERICA'S LIGHTHOUSES: An Illustrated History, Francis Ross Holland, Jr. Delightfully written, profusely illustrated fact-filled survey of over 200 American lighthouses since 1716. History, anecdotes, technological advances, more. 240pp. 8 x 10¾. 25576-X

TOWARDS A NEW ARCHITECTURE, Le Corbusier. Pioneering manifesto by founder of "International School." Technical and aesthetic theories, views of industry, economics, relation of form to function, "mass-production split" and much more. Profusely illustrated. 320pp. 6⅛ x 9¼. (Available in U.S. only.) 25023-7

HOW THE OTHER HALF LIVES, Jacob Riis. Famous journalistic record, exposing poverty and degradation of New York slums around 1900, by major social reformer. 100 striking and influential photographs. 233pp. 10 x 7⅞. 22012-5

FRUIT KEY AND TWIG KEY TO TREES AND SHRUBS, William M. Harlow. One of the handiest and most widely used identification aids. Fruit key covers 120 deciduous and evergreen species; twig key 160 deciduous species. Easily used. Over 300 photographs. 126pp. 5⅜ x 8½. 20511-8

COMMON BIRD SONGS, Dr. Donald J. Borror. Songs of 60 most common U.S. birds: robins, sparrows, cardinals, bluejays, finches, more—arranged in order of increasing complexity. Up to 9 variations of songs of each species.
Cassette and manual 99911-4

ORCHIDS AS HOUSE PLANTS, Rebecca Tyson Northen. Grow cattleyas and many other kinds of orchids—in a window, in a case, or under artificial light. 63 illustrations. 148pp. 5⅜ x 8½. 23261-1

MONSTER MAZES, Dave Phillips. Masterful mazes at four levels of difficulty. Avoid deadly perils and evil creatures to find magical treasures. Solutions for all 32 exciting illustrated puzzles. 48pp. 8¼ x 11. 26005-4

MOZART'S DON GIOVANNI (DOVER OPERA LIBRETTO SERIES), Wolfgang Amadeus Mozart. Introduced and translated by Ellen H. Bleiler. Standard Italian libretto, with complete English translation. Convenient and thoroughly portable—an ideal companion for reading along with a recording or the performance itself. Introduction. List of characters. Plot summary. 121pp. 5¼ x 8½. 24944-1

TECHNICAL MANUAL AND DICTIONARY OF CLASSICAL BALLET, Gail Grant. Defines, explains, comments on steps, movements, poses and concepts. 15-page pictorial section. Basic book for student, viewer. 127pp. 5⅜ x 8½. 21843-0

THE CLARINET AND CLARINET PLAYING, David Pino. Lively, comprehensive work features suggestions about technique, musicianship, and musical interpretation, as well as guidelines for teaching, making your own reeds, and preparing for public performance. Includes an intriguing look at clarinet history. "A godsend," *The Clarinet,* Journal of the International Clarinet Society. Appendixes. 7 illus. 320pp. 5⅜ x 8½. 40270-3

HOLLYWOOD GLAMOR PORTRAITS, John Kobal (ed.). 145 photos from 1926-49. Harlow, Gable, Bogart, Bacall; 94 stars in all. Full background on photographers, technical aspects. 160pp. 8⅜ x 11¼. 23352-9

THE ANNOTATED CASEY AT THE BAT: A Collection of Ballads about the Mighty Casey/Third, Revised Edition, Martin Gardner (ed.). Amusing sequels and parodies of one of America's best-loved poems: Casey's Revenge, Why Casey Whiffed, Casey's Sister at the Bat, others. 256pp. 5⅜ x 8½. 28598-7

THE RAVEN AND OTHER FAVORITE POEMS, Edgar Allan Poe. Over 40 of the author's most memorable poems: "The Bells," "Ulalume," "Israfel," "To Helen," "The Conqueror Worm," "Eldorado," "Annabel Lee," many more. Alphabetic lists of titles and first lines. 64pp. 5³⁄₁₆ x 8¼. 26685-0

PERSONAL MEMOIRS OF U. S. GRANT, Ulysses Simpson Grant. Intelligent, deeply moving firsthand account of Civil War campaigns, considered by many the finest military memoirs ever written. Includes letters, historic photographs, maps and more. 528pp. 6⅛ x 9¼. 28587-1

ANCIENT EGYPTIAN MATERIALS AND INDUSTRIES, A. Lucas and J. Harris. Fascinating, comprehensive, thoroughly documented text describes this ancient civilization's vast resources and the processes that incorporated them in daily life, including the use of animal products, building materials, cosmetics, perfumes and incense, fibers, glazed ware, glass and its manufacture, materials used in the mummification process, and much more. 544pp. 6⅛ x 9¼. (Available in U.S. only.) 40446-3

RUSSIAN STORIES/RUSSKIE RASSKAZY: A Dual-Language Book, edited by Gleb Struve. Twelve tales by such masters as Chekhov, Tolstoy, Dostoevsky, Pushkin, others. Excellent word-for-word English translations on facing pages, plus teaching and study aids, Russian/English vocabulary, biographical/critical introductions, more. 416pp. 5⅜ x 8½. 26244-8

PHILADELPHIA THEN AND NOW: 60 Sites Photographed in the Past and Present, Kenneth Finkel and Susan Oyama. Rare photographs of City Hall, Logan Square, Independence Hall, Betsy Ross House, other landmarks juxtaposed with contemporary views. Captures changing face of historic city. Introduction. Captions. 128pp. 8¼ x 11. 25790-8

AIA ARCHITECTURAL GUIDE TO NASSAU AND SUFFOLK COUNTIES, LONG ISLAND, The American Institute of Architects, Long Island Chapter, and the Society for the Preservation of Long Island Antiquities. Comprehensive, well-researched and generously illustrated volume brings to life over three centuries of Long Island's great architectural heritage. More than 240 photographs with authoritative, extensively detailed captions. 176pp. 8¼ x 11. 26946-9

NORTH AMERICAN INDIAN LIFE: Customs and Traditions of 23 Tribes, Elsie Clews Parsons (ed.). 27 fictionalized essays by noted anthropologists examine religion, customs, government, additional facets of life among the Winnebago, Crow, Zuni, Eskimo, other tribes. 480pp. 6⅛ x 9¼. 27377-6

CATALOG OF DOVER BOOKS

FRANK LLOYD WRIGHT'S DANA HOUSE, Donald Hoffmann. Pictorial essay of residential masterpiece with over 160 interior and exterior photos, plans, elevations, sketches and studies. 128pp. $9^{1}/_{4}$ x $10^{3}/_{4}$. 29120-0

THE MALE AND FEMALE FIGURE IN MOTION: 60 Classic Photographic Sequences, Eadweard Muybridge. 60 true-action photographs of men and women walking, running, climbing, bending, turning, etc., reproduced from rare 19th-century masterpiece. vi + 121pp. 9 x 12. 24745-7

1001 QUESTIONS ANSWERED ABOUT THE SEASHORE, N. J. Berrill and Jacquelyn Berrill. Queries answered about dolphins, sea snails, sponges, starfish, fishes, shore birds, many others. Covers appearance, breeding, growth, feeding, much more. 305pp. $5^{1}/_{4}$ x $8^{1}/_{4}$. 23366-9

ATTRACTING BIRDS TO YOUR YARD, William J. Weber. Easy-to-follow guide offers advice on how to attract the greatest diversity of birds: birdhouses, feeders, water and waterers, much more. 96pp. $5^{3}/_{16}$ x $8^{1}/_{4}$. 28927-3

MEDICINAL AND OTHER USES OF NORTH AMERICAN PLANTS: A Historical Survey with Special Reference to the Eastern Indian Tribes, Charlotte Erichsen-Brown. Chronological historical citations document 500 years of usage of plants, trees, shrubs native to eastern Canada, northeastern U.S. Also complete identifying information. 343 illustrations. 544pp. $6^{1}/_{2}$ x $9^{1}/_{4}$. 25951-X

STORYBOOK MAZES, Dave Phillips. 23 stories and mazes on two-page spreads: Wizard of Oz, Treasure Island, Robin Hood, etc. Solutions. 64pp. $8^{1}/_{4}$ x 11. 23628-5

AMERICAN NEGRO SONGS: 230 Folk Songs and Spirituals, Religious and Secular, John W. Work. This authoritative study traces the African influences of songs sung and played by black Americans at work, in church, and as entertainment. The author discusses the lyric significance of such songs as "Swing Low, Sweet Chariot," "John Henry," and others and offers the words and music for 230 songs. Bibliography. Index of Song Titles. 272pp. $6^{1}/_{2}$ x $9^{1}/_{4}$. 40271-1

MOVIE-STAR PORTRAITS OF THE FORTIES, John Kobal (ed.). 163 glamor, studio photos of 106 stars of the 1940s: Rita Hayworth, Ava Gardner, Marlon Brando, Clark Gable, many more. 176pp. $8^{3}/_{8}$ x $11^{1}/_{4}$. 23546-7

BENCHLEY LOST AND FOUND, Robert Benchley. Finest humor from early 30s, about pet peeves, child psychologists, post office and others. Mostly unavailable elsewhere. 73 illustrations by Peter Arno and others. 183pp. $5^{3}/_{8}$ x $8^{1}/_{2}$. 22410-4

YEKL and THE IMPORTED BRIDEGROOM AND OTHER STORIES OF YIDDISH NEW YORK, Abraham Cahan. Film Hester Street based on *Yekl* (1896). Novel, other stories among first about Jewish immigrants on N.Y.'s East Side. 240pp. $5^{3}/_{8}$ x $8^{1}/_{2}$. 22427-9

SELECTED POEMS, Walt Whitman. Generous sampling from *Leaves of Grass*. Twenty-four poems include "I Hear America Singing," "Song of the Open Road," "I Sing the Body Electric," "When Lilacs Last in the Dooryard Bloom'd," "O Captain! My Captain!"–all reprinted from an authoritative edition. Lists of titles and first lines. 128pp. $5^{3}/_{16}$ x $8^{1}/_{4}$. 26878-0

CATALOG OF DOVER BOOKS

THE BEST TALES OF HOFFMANN, E. T. A. Hoffmann. 10 of Hoffmann's most important stories: "Nutcracker and the King of Mice," "The Golden Flowerpot," etc. 458pp. 5⅜ x 8½. 21793-0

FROM FETISH TO GOD IN ANCIENT EGYPT, E. A. Wallis Budge. Rich detailed survey of Egyptian conception of "God" and gods, magic, cult of animals, Osiris, more. Also, superb English translations of hymns and legends. 240 illustrations. 545pp. 5⅜ x 8½. 25803-3

FRENCH STORIES/CONTES FRANÇAIS: A Dual-Language Book, Wallace Fowlie. Ten stories by French masters, Voltaire to Camus: "Micromegas" by Voltaire; "The Atheist's Mass" by Balzac; "Minuet" by de Maupassant; "The Guest" by Camus, six more. Excellent English translations on facing pages. Also French-English vocabulary list, exercises, more. 352pp. 5⅜ x 8½. 26443-2

CHICAGO AT THE TURN OF THE CENTURY IN PHOTOGRAPHS: 122 Historic Views from the Collections of the Chicago Historical Society, Larry A. Viskochil. Rare large-format prints offer detailed views of City Hall, State Street, the Loop, Hull House, Union Station, many other landmarks, circa 1904-1913. Introduction. Captions. Maps. 144pp. 9⅜ x 12¼. 24656-6

OLD BROOKLYN IN EARLY PHOTOGRAPHS, 1865-1929, William Lee Younger. Luna Park, Gravesend race track, construction of Grand Army Plaza, moving of Hotel Brighton, etc. 157 previously unpublished photographs. 165pp. 8⅞ x 11¾. 23587-4

THE MYTHS OF THE NORTH AMERICAN INDIANS, Lewis Spence. Rich anthology of the myths and legends of the Algonquins, Iroquois, Pawnees and Sioux, prefaced by an extensive historical and ethnological commentary. 36 illustrations. 480pp. 5⅜ x 8½. 25967-6

AN ENCYCLOPEDIA OF BATTLES: Accounts of Over 1,560 Battles from 1479 B.C. to the Present, David Eggenberger. Essential details of every major battle in recorded history from the first battle of Megiddo in 1479 B.C. to Grenada in 1984. List of Battle Maps. New Appendix covering the years 1967-1984. Index. 99 illustrations. 544pp. 6½ x 9¼. 24913-1

SAILING ALONE AROUND THE WORLD, Captain Joshua Slocum. First man to sail around the world, alone, in small boat. One of great feats of seamanship told in delightful manner. 67 illustrations. 294pp. 5⅜ x 8½. 20326-3

ANARCHISM AND OTHER ESSAYS, Emma Goldman. Powerful, penetrating, prophetic essays on direct action, role of minorities, prison reform, puritan hypocrisy, violence, etc. 271pp. 5⅜ x 8½. 22484-8

MYTHS OF THE HINDUS AND BUDDHISTS, Ananda K. Coomaraswamy and Sister Nivedita. Great stories of the epics; deeds of Krishna, Shiva, taken from puranas, Vedas, folk tales; etc. 32 illustrations. 400pp. 5⅜ x 8½. 21759-0

THE TRAUMA OF BIRTH, Otto Rank. Rank's controversial thesis that anxiety neurosis is caused by profound psychological trauma which occurs at birth. 256pp. 5⅜ x 8½. 27974-X

A THEOLOGICO-POLITICAL TREATISE, Benedict Spinoza. Also contains unfinished Political Treatise. Great classic on religious liberty, theory of government on common consent. R. Elwes translation. Total of 421pp. 5⅜ x 8½. 20249-6

MY BONDAGE AND MY FREEDOM, Frederick Douglass. Born a slave, Douglass became outspoken force in antislavery movement. The best of Douglass' autobiographies. Graphic description of slave life. 464pp. 5⅜ x 8½. 22457-0

FOLLOWING THE EQUATOR: A Journey Around the World, Mark Twain. Fascinating humorous account of 1897 voyage to Hawaii, Australia, India, New Zealand, etc. Ironic, bemused reports on peoples, customs, climate, flora and fauna, politics, much more. 197 illustrations. 720pp. 5⅜ x 8½. 26113-1

THE PEOPLE CALLED SHAKERS, Edward D. Andrews. Definitive study of Shakers: origins, beliefs, practices, dances, social organization, furniture and crafts, etc. 33 illustrations. 351pp. 5⅜ x 8½. 21081-2

THE MYTHS OF GREECE AND ROME, H. A. Guerber. A classic of mythology, generously illustrated, long prized for its simple, graphic, accurate retelling of the principal myths of Greece and Rome, and for its commentary on their origins and significance. With 64 illustrations by Michelangelo, Raphael, Titian, Rubens, Canova, Bernini and others. 480pp. 5⅜ x 8½. 27584-1

PSYCHOLOGY OF MUSIC, Carl E. Seashore. Classic work discusses music as a medium from psychological viewpoint. Clear treatment of physical acoustics, auditory apparatus, sound perception, development of musical skills, nature of musical feeling, host of other topics. 88 figures. 408pp. 5⅜ x 8½. 21851-1

THE PHILOSOPHY OF HISTORY, Georg W. Hegel. Great classic of Western thought develops concept that history is not chance but rational process, the evolution of freedom. 457pp. 5⅜ x 8½. 20112-0

THE BOOK OF TEA, Kakuzo Okakura. Minor classic of the Orient: entertaining, charming explanation, interpretation of traditional Japanese culture in terms of tea ceremony. 94pp. 5⅜ x 8½. 20070-1

LIFE IN ANCIENT EGYPT, Adolf Erman. Fullest, most thorough, detailed older account with much not in more recent books, domestic life, religion, magic, medicine, commerce, much more. Many illustrations reproduce tomb paintings, carvings, hieroglyphs, etc. 597pp. 5⅜ x 8½. 22632-8

SUNDIALS, Their Theory and Construction, Albert Waugh. Far and away the best, most thorough coverage of ideas, mathematics concerned, types, construction, adjusting anywhere. Simple, nontechnical treatment allows even children to build several of these dials. Over 100 illustrations. 230pp. 5⅜ x 8½. 22947-5

THEORETICAL HYDRODYNAMICS, L. M. Milne-Thomson. Classic exposition of the mathematical theory of fluid motion, applicable to both hydrodynamics and aerodynamics. Over 600 exercises. 768pp. 6⅛ x 9¼. 68970-0

SONGS OF EXPERIENCE: Facsimile Reproduction with 26 Plates in Full Color, William Blake. 26 full-color plates from a rare 1826 edition. Includes "The Tyger," "London," "Holy Thursday," and other poems. Printed text of poems. 48pp. 5¼ x 7. 24636-1

OLD-TIME VIGNETTES IN FULL COLOR, Carol Belanger Grafton (ed.). Over 390 charming, often sentimental illustrations, selected from archives of Victorian graphics–pretty women posing, children playing, food, flowers, kittens and puppies, smiling cherubs, birds and butterflies, much more. All copyright-free. 48pp. 9¼ x 12¼. 27269-9

PERSPECTIVE FOR ARTISTS, Rex Vicat Cole. Depth, perspective of sky and sea, shadows, much more, not usually covered. 391 diagrams, 81 reproductions of drawings and paintings. 279pp. 5⅜ x 8½. 22487-2

DRAWING THE LIVING FIGURE, Joseph Sheppard. Innovative approach to artistic anatomy focuses on specifics of surface anatomy, rather than muscles and bones. Over 170 drawings of live models in front, back and side views, and in widely varying poses. Accompanying diagrams. 177 illustrations. Introduction. Index. 144pp. 8⅜ x11¼. 26723-7

GOTHIC AND OLD ENGLISH ALPHABETS: 100 Complete Fonts, Dan X. Solo. Add power, elegance to posters, signs, other graphics with 100 stunning copyright-free alphabets: Blackstone, Dolbey, Germania, 97 more–including many lower-case, numerals, punctuation marks. 104pp. 8⅛ x 11. 24695-7

HOW TO DO BEADWORK, Mary White. Fundamental book on craft from simple projects to five-bead chains and woven works. 106 illustrations. 142pp. 5⅜ x 8. 20697-1

THE BOOK OF WOOD CARVING, Charles Marshall Sayers. Finest book for beginners discusses fundamentals and offers 34 designs. "Absolutely first rate . . . well thought out and well executed."–E. J. Tangerman. 118pp. 7¾ x 10⅝. 23654-4

ILLUSTRATED CATALOG OF CIVIL WAR MILITARY GOODS: Union Army Weapons, Insignia, Uniform Accessories, and Other Equipment, Schuyler, Hartley, and Graham. Rare, profusely illustrated 1846 catalog includes Union Army uniform and dress regulations, arms and ammunition, coats, insignia, flags, swords, rifles, etc. 226 illustrations. 160pp. 9 x 12. 24939-5

WOMEN'S FASHIONS OF THE EARLY 1900s: An Unabridged Republication of "New York Fashions, 1909," National Cloak & Suit Co. Rare catalog of mail-order fashions documents women's and children's clothing styles shortly after the turn of the century. Captions offer full descriptions, prices. Invaluable resource for fashion, costume historians. Approximately 725 illustrations. 128pp. 8⅜ x 11¼. 27276-1

THE 1912 AND 1915 GUSTAV STICKLEY FURNITURE CATALOGS, Gustav Stickley. With over 200 detailed illustrations and descriptions, these two catalogs are essential reading and reference materials and identification guides for Stickley furniture. Captions cite materials, dimensions and prices. 112pp. 6½ x 9¼. 26676-1

EARLY AMERICAN LOCOMOTIVES, John H. White, Jr. Finest locomotive engravings from early 19th century: historical (1804–74), main-line (after 1870), special, foreign, etc. 147 plates. 142pp. 11⅜ x 8¼. 22772-3

THE TALL SHIPS OF TODAY IN PHOTOGRAPHS, Frank O. Braynard. Lavishly illustrated tribute to nearly 100 majestic contemporary sailing vessels: Amerigo Vespucci, Clearwater, Constitution, Eagle, Mayflower, Sea Cloud, Victory, many more. Authoritative captions provide statistics, background on each ship. 190 black-and-white photographs and illustrations. Introduction. 128pp. 8⅞ x 11¾. 27163-3

LITTLE BOOK OF EARLY AMERICAN CRAFTS AND TRADES, Peter Stockham (ed.). 1807 children's book explains crafts and trades: baker, hatter, cooper, potter, and many others. 23 copperplate illustrations. 140pp. $4^5/_8$ x 6. 23336-7

VICTORIAN FASHIONS AND COSTUMES FROM HARPER'S BAZAR, 1867–1898, Stella Blum (ed.). Day costumes, evening wear, sports clothes, shoes, hats, other accessories in over 1,000 detailed engravings. 320pp. $9\frac{3}{8}$ x $12\frac{1}{4}$. 22990-4

GUSTAV STICKLEY, THE CRAFTSMAN, Mary Ann Smith. Superb study surveys broad scope of Stickley's achievement, especially in architecture. Design philosophy, rise and fall of the Craftsman empire, descriptions and floor plans for many Craftsman houses, more. 86 black-and-white halftones. 31 line illustrations. Introduction 208pp. $6\frac{1}{2}$ x $9\frac{1}{4}$. 27210-9

THE LONG ISLAND RAIL ROAD IN EARLY PHOTOGRAPHS, Ron Ziel. Over 220 rare photos, informative text document origin (1844) and development of rail service on Long Island. Vintage views of early trains, locomotives, stations, passengers, crews, much more. Captions. $8\frac{7}{8}$ x $11\frac{3}{4}$. 26301-0

VOYAGE OF THE LIBERDADE, Joshua Slocum. Great 19th-century mariner's thrilling, first-hand account of the wreck of his ship off South America, the 35-foot boat he built from the wreckage, and its remarkable voyage home. 128pp. $5\frac{3}{8}$ x $8\frac{1}{2}$. 40022-0

TEN BOOKS ON ARCHITECTURE, Vitruvius. The most important book ever written on architecture. Early Roman aesthetics, technology, classical orders, site selection, all other aspects. Morgan translation. 331pp. $5\frac{3}{8}$ x $8\frac{1}{2}$. 20645-9

THE HUMAN FIGURE IN MOTION, Eadweard Muybridge. More than 4,500 stopped-action photos, in action series, showing undraped men, women, children jumping, lying down, throwing, sitting, wrestling, carrying, etc. 390pp. $7\frac{7}{8}$ x $10\frac{5}{8}$. 20204-6 Clothbd.

TREES OF THE EASTERN AND CENTRAL UNITED STATES AND CANADA, William M. Harlow. Best one-volume guide to 140 trees. Full descriptions, woodlore, range, etc. Over 600 illustrations. Handy size. 288pp. $4\frac{1}{2}$ x $6\frac{5}{8}$. 20395-6

SONGS OF WESTERN BIRDS, Dr. Donald J. Borror. Complete song and call repertoire of 60 western species, including flycatchers, juncoes, cactus wrens, many more—includes fully illustrated booklet. Cassette and manual 99913-0

GROWING AND USING HERBS AND SPICES, Milo Miloradovich. Versatile handbook provides all the information needed for cultivation and use of all the herbs and spices available in North America. 4 illustrations. Index. Glossary. 236pp. $5\frac{3}{8}$ x $8\frac{1}{2}$. 25058-X

BIG BOOK OF MAZES AND LABYRINTHS, Walter Shepherd. 50 mazes and labyrinths in all—classical, solid, ripple, and more—in one great volume. Perfect inexpensive puzzler for clever youngsters. Full solutions. 112pp. $8\frac{1}{4}$ x 11. 22951-3

CATALOG OF DOVER BOOKS

PIANO TUNING, J. Cree Fischer. Clearest, best book for beginner, amateur. Simple repairs, raising dropped notes, tuning by easy method of flattened fifths. No previous skills needed. 4 illustrations. 201pp. 5⅜ x 8½. 23267-0

HINTS TO SINGERS, Lillian Nordica. Selecting the right teacher, developing confidence, overcoming stage fright, and many other important skills receive thoughtful discussion in this indispensible guide, written by a world-famous diva of four decades' experience. 96pp. 5⅜ x 8½. 40094-8

THE COMPLETE NONSENSE OF EDWARD LEAR, Edward Lear. All nonsense limericks, zany alphabets, Owl and Pussycat, songs, nonsense botany, etc., illustrated by Lear. Total of 320pp. 5⅜ x 8½. (Available in U.S. only.) 20167-8

VICTORIAN PARLOUR POETRY: An Annotated Anthology, Michael R. Turner. 117 gems by Longfellow, Tennyson, Browning, many lesser-known poets. "The Village Blacksmith," "Curfew Must Not Ring Tonight," "Only a Baby Small," dozens more, often difficult to find elsewhere. Index of poets, titles, first lines. xxiii + 325pp. 5⅜ x 8¼. 27044-0

DUBLINERS, James Joyce. Fifteen stories offer vivid, tightly focused observations of the lives of Dublin's poorer classes. At least one, "The Dead," is considered a masterpiece. Reprinted complete and unabridged from standard edition. 160pp. 5³⁄₁₆ x 8¼. 26870-5

GREAT WEIRD TALES: 14 Stories by Lovecraft, Blackwood, Machen and Others, S. T. Joshi (ed.). 14 spellbinding tales, including "The Sin Eater," by Fiona McLeod, "The Eye Above the Mantel," by Frank Belknap Long, as well as renowned works by R. H. Barlow, Lord Dunsany, Arthur Machen, W. C. Morrow and eight other masters of the genre. 256pp. 5⅜ x 8½. (Available in U.S. only.) 40436-6

THE BOOK OF THE SACRED MAGIC OF ABRAMELIN THE MAGE, translated by S. MacGregor Mathers. Medieval manuscript of ceremonial magic. Basic document in Aleister Crowley, Golden Dawn groups. 268pp. 5⅜ x 8½. 23211-5

NEW RUSSIAN-ENGLISH AND ENGLISH-RUSSIAN DICTIONARY, M. A. O'Brien. This is a remarkably handy Russian dictionary, containing a surprising amount of information, including over 70,000 entries. 366pp. 4½ x 6⅛. 20208-9

HISTORIC HOMES OF THE AMERICAN PRESIDENTS, Second, Revised Edition, Irvin Haas. A traveler's guide to American Presidential homes, most open to the public, depicting and describing homes occupied by every American President from George Washington to George Bush. With visiting hours, admission charges, travel routes. 175 photographs. Index. 160pp. 8¼ x 11. 26751-2

NEW YORK IN THE FORTIES, Andreas Feininger. 162 brilliant photographs by the well-known photographer, formerly with *Life* magazine. Commuters, shoppers, Times Square at night, much else from city at its peak. Captions by John von Hartz. 181pp. 9¼ x 10¾. 23585-8

INDIAN SIGN LANGUAGE, William Tomkins. Over 525 signs developed by Sioux and other tribes. Written instructions and diagrams. Also 290 pictographs. 111pp. 6⅛ x 9¼. 22029-X

CATALOG OF DOVER BOOKS

ANATOMY: A Complete Guide for Artists, Joseph Sheppard. A master of figure drawing shows artists how to render human anatomy convincingly. Over 460 illustrations. 224pp. 8⅜ x 11¼. 27279-6

MEDIEVAL CALLIGRAPHY: Its History and Technique, Marc Drogin. Spirited history, comprehensive instruction manual covers 13 styles (ca. 4th century through 15th). Excellent photographs; directions for duplicating medieval techniques with modern tools. 224pp. 8⅜ x 11¼. 26142-5

DRIED FLOWERS: How to Prepare Them, Sarah Whitlock and Martha Rankin. Complete instructions on how to use silica gel, meal and borax, perlite aggregate, sand and borax, glycerine and water to create attractive permanent flower arrangements. 12 illustrations. 32pp. 5⅜ x 8½. 21802-3

EASY-TO-MAKE BIRD FEEDERS FOR WOODWORKERS, Scott D. Campbell. Detailed, simple-to-use guide for designing, constructing, caring for and using feeders. Text, illustrations for 12 classic and contemporary designs. 96pp. 5⅜ x 8½. 25847-5

SCOTTISH WONDER TALES FROM MYTH AND LEGEND, Donald A. Mackenzie. 16 lively tales tell of giants rumbling down mountainsides, of a magic wand that turns stone pillars into warriors, of gods and goddesses, evil hags, powerful forces and more. 240pp. 5⅜ x 8½. 29677-6

THE HISTORY OF UNDERCLOTHES, C. Willett Cunnington and Phyllis Cunnington. Fascinating, well-documented survey covering six centuries of English undergarments, enhanced with over 100 illustrations: 12th-century laced-up bodice, footed long drawers (1795), 19th-century bustles, 19th-century corsets for men, Victorian "bust improvers," much more. 272pp. 5⅝ x 8¼. 27124-2

ARTS AND CRAFTS FURNITURE: The Complete Brooks Catalog of 1912, Brooks Manufacturing Co. Photos and detailed descriptions of more than 150 now very collectible furniture designs from the Arts and Crafts movement depict davenports, settees, buffets, desks, tables, chairs, bedsteads, dressers and more, all built of solid, quarter-sawed oak. Invaluable for students and enthusiasts of antiques, Americana and the decorative arts. 80pp. 6½ x 9¼. 27471-3

WILBUR AND ORVILLE: A Biography of the Wright Brothers, Fred Howard. Definitive, crisply written study tells the full story of the brothers' lives and work. A vividly written biography, unparalleled in scope and color, that also captures the spirit of an extraordinary era. 560pp. 6⅛ x 9¼. 40297-5

THE ARTS OF THE SAILOR: Knotting, Splicing and Ropework, Hervey Garrett Smith. Indispensable shipboard reference covers tools, basic knots and useful hitches; handsewing and canvas work, more. Over 100 illustrations. Delightful reading for sea lovers. 256pp. 5⅜ x 8½. 26440-8

FRANK LLOYD WRIGHT'S FALLINGWATER: The House and Its History, Second, Revised Edition, Donald Hoffmann. A total revision–both in text and illustrations–of the standard document on Fallingwater, the boldest, most personal architectural statement of Wright's mature years, updated with valuable new material from the recently opened Frank Lloyd Wright Archives. "Fascinating"–*The New York Times*. 116 illustrations. 128pp. 9¼ x 10¾. 27430-6

CATALOG OF DOVER BOOKS

PHOTOGRAPHIC SKETCHBOOK OF THE CIVIL WAR, Alexander Gardner. 100 photos taken on field during the Civil War. Famous shots of Manassas Harper's Ferry, Lincoln, Richmond, slave pens, etc. 244pp. $10\frac{5}{8}$ x $8\frac{1}{4}$. 22731-6

FIVE ACRES AND INDEPENDENCE, Maurice G. Kains. Great back-to-the-land classic explains basics of self-sufficient farming. The one book to get. 95 illustrations. 397pp. $5\frac{3}{8}$ x $8\frac{1}{2}$. 20974-1

SONGS OF EASTERN BIRDS, Dr. Donald J. Borror. Songs and calls of 60 species most common to eastern U.S.: warblers, woodpeckers, flycatchers, thrushes, larks, many more in high-quality recording. Cassette and manual 99912-2

A MODERN HERBAL, Margaret Grieve. Much the fullest, most exact, most useful compilation of herbal material. Gigantic alphabetical encyclopedia, from aconite to zedoary, gives botanical information, medical properties, folklore, economic uses, much else. Indispensable to serious reader. 161 illustrations. 888pp. $6\frac{1}{2}$ x $9\frac{1}{4}$. 2-vol. set. (Available in U.S. only.) Vol. I: 22798-7
Vol. II: 22799-5

HIDDEN TREASURE MAZE BOOK, Dave Phillips. Solve 34 challenging mazes accompanied by heroic tales of adventure. Evil dragons, people-eating plants, bloodthirsty giants, many more dangerous adversaries lurk at every twist and turn. 34 mazes, stories, solutions. 48pp. $8\frac{1}{4}$ x 11. 24566-7

LETTERS OF W. A. MOZART, Wolfgang A. Mozart. Remarkable letters show bawdy wit, humor, imagination, musical insights, contemporary musical world; includes some letters from Leopold Mozart. 276pp. $5\frac{3}{8}$ x $8\frac{1}{2}$. 22859-2

BASIC PRINCIPLES OF CLASSICAL BALLET, Agrippina Vaganova. Great Russian theoretician, teacher explains methods for teaching classical ballet. 118 illustrations. 175pp. $5\frac{3}{8}$ x $8\frac{1}{2}$. 22036-2

THE JUMPING FROG, Mark Twain. Revenge edition. The original story of The Celebrated Jumping Frog of Calaveras County, a hapless French translation, and Twain's hilarious "retranslation" from the French. 12 illustrations. 66pp. $5\frac{3}{8}$ x $8\frac{1}{2}$.
22686-7

BEST REMEMBERED POEMS, Martin Gardner (ed.). The 126 poems in this superb collection of 19th- and 20th-century British and American verse range from Shelley's "To a Skylark" to the impassioned "Renascence" of Edna St. Vincent Millay and to Edward Lear's whimsical "The Owl and the Pussycat." 224pp. $5\frac{3}{8}$ x $8\frac{1}{2}$.
27165-X

COMPLETE SONNETS, William Shakespeare. Over 150 exquisite poems deal with love, friendship, the tyranny of time, beauty's evanescence, death and other themes in language of remarkable power, precision and beauty. Glossary of archaic terms. 80pp. $5\frac{3}{16}$ x $8\frac{1}{4}$. 26686-9

THE BATTLES THAT CHANGED HISTORY, Fletcher Pratt. Eminent historian profiles 16 crucial conflicts, ancient to modern, that changed the course of civilization. 352pp. $5\frac{3}{8}$ x $8\frac{1}{2}$. 41129-X

THE WIT AND HUMOR OF OSCAR WILDE, Alvin Redman (ed.). More than 1,000 ripostes, paradoxes, wisecracks: Work is the curse of the drinking classes; I can resist everything except temptation; etc. 258pp. 5⅜ x 8½. 20602-5

SHAKESPEARE LEXICON AND QUOTATION DICTIONARY, Alexander Schmidt. Full definitions, locations, shades of meaning in every word in plays and poems. More than 50,000 exact quotations. 1,485pp. 6½ x 9¼. 2-vol. set.
Vol. 1: 22726-X
Vol. 2: 22727-8

SELECTED POEMS, Emily Dickinson. Over 100 best-known, best-loved poems by one of America's foremost poets, reprinted from authoritative early editions. No comparable edition at this price. Index of first lines. 64pp. 5³⁄₁₆ x 8¼. 26466-1

THE INSIDIOUS DR. FU-MANCHU, Sax Rohmer. The first of the popular mystery series introduces a pair of English detectives to their archnemesis, the diabolical Dr. Fu-Manchu. Flavorful atmosphere, fast-paced action, and colorful characters enliven this classic of the genre. 208pp. 5³⁄₁₆ x 8¼. 29898-1

THE MALLEUS MALEFICARUM OF KRAMER AND SPRENGER, translated by Montague Summers. Full text of most important witchhunter's "bible," used by both Catholics and Protestants. 278pp. 6⅝ x 10. 22802-9

SPANISH STORIES/CUENTOS ESPAÑOLES: A Dual-Language Book, Angel Flores (ed.). Unique format offers 13 great stories in Spanish by Cervantes, Borges, others. Faithful English translations on facing pages. 352pp. 5⅜ x 8½. 25399-6

GARDEN CITY, LONG ISLAND, IN EARLY PHOTOGRAPHS, 1869–1919, Mildred H. Smith. Handsome treasury of 118 vintage pictures, accompanied by carefully researched captions, document the Garden City Hotel fire (1899), the Vanderbilt Cup Race (1908), the first airmail flight departing from the Nassau Boulevard Aerodrome (1911), and much more. 96pp. 8⅞ x 11¾. 40669-5

OLD QUEENS, N.Y., IN EARLY PHOTOGRAPHS, Vincent F. Seyfried and William Asadorian. Over 160 rare photographs of Maspeth, Jamaica, Jackson Heights, and other areas. Vintage views of DeWitt Clinton mansion, 1939 World's Fair and more. Captions. 192pp. 8⅞ x 11. 26358-4

CAPTURED BY THE INDIANS: 15 Firsthand Accounts, 1750-1870, Frederick Drimmer. Astounding true historical accounts of grisly torture, bloody conflicts, relentless pursuits, miraculous escapes and more, by people who lived to tell the tale. 384pp. 5⅜ x 8½. 24901-8

THE WORLD'S GREAT SPEECHES (Fourth Enlarged Edition), Lewis Copeland, Lawrence W. Lamm, and Stephen J. McKenna. Nearly 300 speeches provide public speakers with a wealth of updated quotes and inspiration–from Pericles' funeral oration and William Jennings Bryan's "Cross of Gold Speech" to Malcolm X's powerful words on the Black Revolution and Earl of Spenser's tribute to his sister, Diana, Princess of Wales. 944pp. 5⅜ x 8⅜. 40903-1

THE BOOK OF THE SWORD, Sir Richard F. Burton. Great Victorian scholar/adventurer's eloquent, erudite history of the "queen of weapons"–from prehistory to early Roman Empire. Evolution and development of early swords, variations (sabre, broadsword, cutlass, scimitar, etc.), much more. 336pp. 6⅛ x 9¼.
25434-8

CATALOG OF DOVER BOOKS

AUTOBIOGRAPHY: The Story of My Experiments with Truth, Mohandas K. Gandhi. Boyhood, legal studies, purification, the growth of the Satyagraha (nonviolent protest) movement. Critical, inspiring work of the man responsible for the freedom of India. 480pp. 5⅜ x 8½. (Available in U.S. only.) 24593-4

CELTIC MYTHS AND LEGENDS, T. W. Rolleston. Masterful retelling of Irish and Welsh stories and tales. Cuchulain, King Arthur, Deirdre, the Grail, many more. First paperback edition. 58 full-page illustrations. 512pp. 5⅜ x 8½. 26507-2

THE PRINCIPLES OF PSYCHOLOGY, William James. Famous long course complete, unabridged. Stream of thought, time perception, memory, experimental methods; great work decades ahead of its time. 94 figures. 1,391pp. 5⅜ x 8½. 2-vol. set.
Vol. I: 20381-6   Vol. II: 20382-4

THE WORLD AS WILL AND REPRESENTATION, Arthur Schopenhauer. Definitive English translation of Schopenhauer's life work, correcting more than 1,000 errors, omissions in earlier translations. Translated by E. F. J. Payne. Total of 1,269pp. 5⅜ x 8½. 2-vol. set. Vol. 1: 21761-2   Vol. 2: 21762-0

MAGIC AND MYSTERY IN TIBET, Madame Alexandra David-Neel. Experiences among lamas, magicians, sages, sorcerers, Bonpa wizards. A true psychic discovery. 32 illustrations. 321pp. 5⅜ x 8½. (Available in U.S. only.) 22682-4

THE EGYPTIAN BOOK OF THE DEAD, E. A. Wallis Budge. Complete reproduction of Ani's papyrus, finest ever found. Full hieroglyphic text, interlinear transliteration, word-for-word translation, smooth translation. 533pp. 6½ x 9¼. 21866-X

MATHEMATICS FOR THE NONMATHEMATICIAN, Morris Kline. Detailed, college-level treatment of mathematics in cultural and historical context, with numerous exercises. Recommended Reading Lists. Tables. Numerous figures. 641pp. 5⅜ x 8½. 24823-2

PROBABILISTIC METHODS IN THE THEORY OF STRUCTURES, Isaac Elishakoff. Well-written introduction covers the elements of the theory of probability from two or more random variables, the reliability of such multivariable structures, the theory of random function, Monte Carlo methods of treating problems incapable of exact solution, and more. Examples. 502pp. 5⅜ x 8½. 40691-1

THE RIME OF THE ANCIENT MARINER, Gustave Doré, S. T. Coleridge. Doré's finest work; 34 plates capture moods, subtleties of poem. Flawless full-size reproductions printed on facing pages with authoritative text of poem. "Beautiful. Simply beautiful."–Publisher's Weekly. 77pp. 9¼ x 12. 22305-1

NORTH AMERICAN INDIAN DESIGNS FOR ARTISTS AND CRAFTSPEOPLE, Eva Wilson. Over 360 authentic copyright-free designs adapted from Navajo blankets, Hopi pottery, Sioux buffalo hides, more. Geometrics, symbolic figures, plant and animal motifs, etc. 128pp. 8⅜ x 11. (Not for sale in the United Kingdom.) 25341-4

SCULPTURE: Principles and Practice, Louis Slobodkin. Step-by-step approach to clay, plaster, metals, stone; classical and modern. 253 drawings, photos. 255pp. 8⅛ x 11. 22960-2

THE INFLUENCE OF SEA POWER UPON HISTORY, 1660–1783, A. T. Mahan. Influential classic of naval history and tactics still used as text in war colleges. First paperback edition. 4 maps. 24 battle plans. 640pp. 5⅜ x 8½. 25509-3

CATALOG OF DOVER BOOKS

THE STORY OF THE TITANIC AS TOLD BY ITS SURVIVORS, Jack Winocour (ed.). What it was really like. Panic, despair, shocking inefficiency, and a little heroism. More thrilling than any fictional account. 26 illustrations. 320pp. 5⅜ x 8½.
20610-6

FAIRY AND FOLK TALES OF THE IRISH PEASANTRY, William Butler Yeats (ed.). Treasury of 64 tales from the twilight world of Celtic myth and legend: "The Soul Cages," "The Kildare Pooka," "King O'Toole and his Goose," many more. Introduction and Notes by W. B. Yeats. 352pp. 5⅜ x 8½.
26941-8

BUDDHIST MAHAYANA TEXTS, E. B. Cowell and others (eds.). Superb, accurate translations of basic documents in Mahayana Buddhism, highly important in history of religions. The Buddha-karita of Asvaghosha, Larger Sukhavativyuha, more. 448pp. 5⅜ x 8½.
25552-2

ONE TWO THREE . . . INFINITY: Facts and Speculations of Science, George Gamow. Great physicist's fascinating, readable overview of contemporary science: number theory, relativity, fourth dimension, entropy, genes, atomic structure, much more. 128 illustrations. Index. 352pp. 5⅜ x 8½.
25664-2

EXPERIMENTATION AND MEASUREMENT, W. J. Youden. Introductory manual explains laws of measurement in simple terms and offers tips for achieving accuracy and minimizing errors. Mathematics of measurement, use of instruments, experimenting with machines. 1994 edition. Foreword. Preface. Introduction. Epilogue. Selected Readings. Glossary. Index. Tables and figures. 128pp. 5⅜ x 8½.
40451-X

DALÍ ON MODERN ART: The Cuckolds of Antiquated Modern Art, Salvador Dalí. Influential painter skewers modern art and its practitioners. Outrageous evaluations of Picasso, Cézanne, Turner, more. 15 renderings of paintings discussed. 44 calligraphic decorations by Dalí. 96pp. 5⅜ x 8½. (Available in U.S. only.)
29220-7

ANTIQUE PLAYING CARDS: A Pictorial History, Henry René D'Allemagne. Over 900 elaborate, decorative images from rare playing cards (14th–20th centuries): Bacchus, death, dancing dogs, hunting scenes, royal coats of arms, players cheating, much more. 96pp. 9¼ x 12¼.
29265-7

MAKING FURNITURE MASTERPIECES: 30 Projects with Measured Drawings, Franklin H. Gottshall. Step-by-step instructions, illustrations for constructing handsome, useful pieces, among them a Sheraton desk, Chippendale chair, Spanish desk, Queen Anne table and a William and Mary dressing mirror. 224pp. 8⅛ x 11¼.
29338-6

THE FOSSIL BOOK: A Record of Prehistoric Life, Patricia V. Rich et al. Profusely illustrated definitive guide covers everything from single-celled organisms and dinosaurs to birds and mammals and the interplay between climate and man. Over 1,500 illustrations. 760pp. 7½ x 10⅛.
29371-8

Paperbound unless otherwise indicated. Available at your book dealer, online at **www.doverpublications.com**, or by writing to Dept. GI, Dover Publications, Inc., 31 East 2nd Street, Mineola, NY 11501. For current price information or for free catalogues (please indicate field of interest), write to Dover Publications or log on to **www.doverpublications.com** and see every Dover book in print. Dover publishes more than 500 books each year on science, elementary and advanced mathematics, biology, music, art, literary history, social sciences, and other areas.